PDA Robotics

PDA Robotics

Using Your Personal Digital Assistant to Control Your Robot

Douglas H. Williams

McGraw-Hill

New York   Chicago   San Francisco   Lisbon   London   Madrid
Mexico City   Milan   New Delhi   San Juan   Seoul
Singapore   Sydney   Toronto

The *McGraw·Hill* Companies

Cataloging-in-Publication Data is on file with the Library of Congress

1 2 3 4 5 6 7 8 9 0   DOC/DOC   0 9 8 7 6 5 4 3

ISBN 0-07-141741-9

The sponsoring editor for this book was Judy Bass and the production supervisor was Sherri Souffrance. It was set in Melior by Patricia Wallenburg.

Printed and bound by RR Donnelley.

 This book was printed on recycled, acid-free paper containing a minimum of 50% recycled, de-inked fiber.

McGraw-Hill books are available at special quantity discounts to use as premiums and sales promotions, or for use in corporate training programs. For more information, please write to the Director of Special Sales, Professional Publishing, McGraw-Hill, Two Penn Plaza, New York, NY 10121-2298. Or contact your local bookstore.

Dedicated to my family, Gylian, Olivia, Rachel, and Ethan.

Contents Summary

Contents

Introduction

The NASA Mars Sojourner rover inspired this project (http://mars.jpl.nasa.gov/MPF/index1.html). I followed the mission with great enthusiasm and witnessed a giant leap in robotics that day it began roaming the Martian terrain and sending images back to earth. Though I was in awe when the Viking missions of the 1970's were in progress, we didn't see that near real-time interaction with the craft (http://nssdc.gsfc.nasa.gov/planetary/viking.html). The twin rovers scheduled to launch May/July 2003 and land on the surface January 2004 will be something to follow (http://mars.jpl.nasa.gov/mer/)! PDA Robot is a scaled down version of Sojourner that has a similar framework, components, and functionality at a much lower cost!

The personal digital assistant is the main control unit of the robot, communicating with the craft's body via a beam of infrared light and to other machines on the wireless network. The PDA itself becomes a data transponder. It (the PDA) is insulated and protected from the robotic interface. It is said to be optically isolated, communicating on ripples of light. Because of this design, no connectors are required and the software provided will work with any Windows or PalmOS driven handheld PDA. I see a day when all components of a system are connectionless with harmonically synchronized transistors.

I will go into the theory behind the operation of each component as well as the practical hands-on information and processes needed to complete this project. I will also make suggestions for enhancements and modifications to the electronic, mechanical, and software design; enhancements that I will leave up to you to explore.

The only limit to any enhancements or changes will be that of your imagination. This book will give you the expertise to create anything. One of many areas that I will touch on is the smart distributed network, where each robot can pass the information that it gains onto the "collective" to be shared with other robots. For instance, if two PDA Robots pass each other they can exchange information about a room in the house that has been mapped, saving any duplication of effort. The robots can synchronize to coordinate effort as well. A good example of a coordinated autonomous effort is the idea of traffic being directed by a computer system. In the future, I believe the key to making the world a better place is to effectively and fully use the resources we have available. Traffic congestion on the freeways could be eliminated for years to come without building anymore highways if it was managed properly. Cars outfitted with sensors and wireless technology could be tied into a central coordination system making the commute to work an enjoyable and relaxing experience. This is something that could be achieved on a smaller scale with this project if you take it a step further.

Artificial intelligence, self-modifying code, and the emergent behavior of computers is a fascinating area of research that will be touched on in this book. Emergent behavior in a system is the system's ability to become intelligent over and above the programming that has been coded into it. Sometime this is seen as a behavior or unanticipated function that is the result of the interaction between two systems. I have seen this happen with smart digital imaging archiving software at the medical imaging company where I currently work. One must be careful when enabling a machine with AI to make decisions around humans though. A "smart" robot building an office tower may decide that the best course of action may be to remove a support beam and put it up at a later time. But if the programmer made a mistake and didn't have another algorithm check the structural integrity before approving of the decision, then the whole building would come down. A simple coding error of "if (StructuralIntegrityOk = TRUE){ RemoveBeam(BeamNumber); }" spells disaster. The equality operator == is mistaken for and the assignment operator =. One must ensure that AI bots stay within safe operating parameters, are monitored closely, and have a remote kill switch.

Enabling the machine with a sense of sight is another topic that will be explored and explained. PDA Robot can "see" through the use of an infrared range finder and wireless video camera. The machine vision algorithms used in this project interpret the surroundings and send feedback to the robot. The ability to send video data into the wireless network through a video capture card open the "window" to a virtual presence. Amazing things are being done today with this technology. Doctors can perform surgery from any point on earth to another; we can be there from here!

One interesting point about the IR range detector is the fact that the pulsed beam of IR light is highly visible to a modern IR target locking system deployed by most modern military equipment. This could be an advantage or a drawback. The invisible infrared beam can provide a good source for a night vision video camera, in fact most low cost video cameras will be able to detect the beam from the front. If you have a video camera give it a try! I will discuss other methods of data transmission (visible light) and range finding (invisible). If we tap into the this range finder and pulse the light beam and use a telescope, we can create a very long range point-to-point communication device ideal for ground to air operations. Something I will leave you to experiment with.

Once PDA Robot is on the network it is essentially an internet appliance.

My hope is that this project will give you the knowledge and experience to create any electronic device that you can dream up. All the information is out there—just follow the links from a good search engine. Automation, ordering over the Web, and courier service allows everything in this project to be delivered to your door. Please experiment with the design—I've designed an amphibious and airborne body that the circuitry can be "snapped" into. I hope you evolve this design once you become familiar with it.

If this technology is applied in the same spirit as the space program and with the ethics of modern medicine, then I can see great things evolving from it

For online updates, source code, and other useful files that will aid you in completing PDA Robot, please visit www.pda-robotics.com.

Douglas Williams

Acknowledgments

Thanks goes out to everyone along the way made this book possible, especially my brothers, Karl Williams and Geoff Williams, whom without I would have not endeavored to write this book. Thanks to my parents, Gord and Ruth Williams, for all their support over the years. Thanks to my family for putting up with my late nights and lost weekends.

Thanks to Judy Bass and Patricia Wallenburg, for their patience and the fabulous job they have done putting the whole thing together.

Special thanks to my friends and colleagues who have inspired me along the way: Michael Foote, Bob Lazic, Paul Stienbach, Dave Huson, Dave Smith, Stephane MacMaster, John Lammers, Julius Avelar, Erkan Akyuz, Desh Sharma, Tim Jones, Tom Cloutier, Paul McNally, Barry Reville, Bart Domzy, James Chase, Stephen Kingston, John Sanio, Kim Martin, Clark MacDonald, Peter Madziak Stephen Frederick, Derrick Barnes, Darren Tarachan, Steve Spicer, Mathew Sullivan, John Kominar, Grant E, Paul Barton, Eric Peterson, Larry Williamson, and anyone I may have left off of this list.

Thanks to Rebecca Tollen for the information on telesurgery and Microsoft, Palm OS, MicroChip, HVW Tech, Sharp, ST Microelectronics, Micro Engineering Labs, Protel, Intel, Intuitive Surgical, Handspring, HP, and Compaq for helping to make this project possible.

PDA Robotics

1

Anatomy of a Personal Digital Assistant (PDA)

The power is sitting in the palm of your hand. The technology exists today to bring your world to you wherever you happen to be. Wireless technology, a handful of electronic components, a small handheld computer, and little software to glue everything together is all that is needed to be "virtually" enabled. The culmination of this project will provide you with the know-how to create a robotic device that can be controlled through your PDA from anywhere over the World Wide Web or allowed to roam autonomously using its PDA "brain."

Why use a PDA? These devices are small and powerful, leveraging the best technology that can be offered today in the palm of your hand. They make for perfect robotic controllers, as they can be easily expanded through their expansion slots. If you need a wireless network or a global positioning system, simply slide in the card. Increasingly, they have the wireless technology built into them, such as Bluetooth or digital/analog cellular phone technology, as seen in **Figure 1.1**. These devices have rich application programming interfaces (APIs) that can be used to create powerful end user applications, capitalizing on the device capabilities, as shown in this book. The Infrared Data Association (IrDA) functions contained in both the Windows CE and Palm OS APIs are pure abstractions to the actual infrared transceivers built into the PDA. For example, socket (AF_IRDA, SOCK_STREAM, NULL) and IrOpen (irref, irOpenOptSpeed115200) are the Windows CE and Palm OS API calls used to initiate the IrDA Data link to the PDA

Figure 1.1

Integrated wireless
PDAs.

Robot. The source included will explain in detail how to accomplish a data link used to send and receive information.

Once the link is established, users can virtually project themselves anywhere. A doctor can perform surgery on a patient thousands of miles away. You can roam around your house on PDA Robot from your hotel room, cottage, or even flying 60,000 feet above the earth. This book will give you the tools and know-how to transform this project into anything. Explaining the schematic design, circuit board manufacturing, embedded software for the microchip, mechanical design and the software source code for the world's two most popular PDA (handheld) operating systems, this book will take you on a tour of today's specialized electronic microchips and the inner workings of PDA operating systems.

PDA (personal digital assistant) is a term for any small mobile handheld device that provides computing and information storage and retrieval capabilities for personal or business use, often for keeping schedule calendars and address book information handy. The term handheld is a synonym. Many people use the name of one of the popular PDA products as a generic term. These include Compaq/Hewlett-Packard's IPAQ and 3Com's Palm devices, such as the Palm Pilot and m505.

Most PDAs have a small keyboard that the PDA clips onto, and an electronically sensitive pad on which handwriting can be received. Typical uses include schedule and address book storage and retrieval and note-entering. However, many applications have been written for PDAs. Increasingly, PDAs are combined with telephones, paging systems, and wireless networks.

Some PDAs offer a variation of the Microsoft Windows operating system called Windows CE (Pocket PC), which offers the familiar "MS Windows" look and feel. Other products, such as the palm devices, have their own operating system called Palm OS.

- *Windows CE:* Windows CE is a Microsoft operating system for handhelds, TV set-top boxes, upcoming home appliances, even game consoles (the new Sega Dreamcast is WinCE compatible). Pocket PCs use Windows CE. Windows CE uses the familiar Windows task bar, scroll bar, and drop-down menus. Unlike Palm devices, WinCE products usually have a color screen.

- *Palm OS:* The Palm operating system runs the Palm series of organizers, the IBM Wordpad series, the new Visor products, and Sony Clie. Palm OS is known for its speedy navigation when compared with Pocket PCs.

- *Pocket PC:* Pocket PCs are a direct competitor to Palm handhelds. They use the Windows CE operating system and have color screens, among other standard features.

Most PDAs are able to communicate directly with each other through the use of an infrared (IR) port. This makes sharing information effortless. By simply lining up IR ports, people can "beam" information back and forth. Documents can be "beamed" directly to a printer or information exchanged bi-directionally to an IR transponder connected to a network.

Many university campuses, such as the University of California at Berkeley, are IR enabled. Students can get class schedules and notes, receive and transmit assignments, and even have the bus schedule beamed directly to them from IR transponders placed around the campus.

The PDA Robot featured in this book will use the IR port on the PDA to communicate with its body. This protects the PDA from any dam-

Figure 1.2

Palm m505: A
typical PDA.

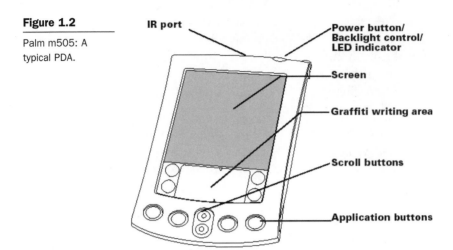

age that could occur by an electronic malfunction in the robot body, and eliminates the need for any physical connection to the PDA. The PDA will act as the "brain" of the robot, monitoring and controlling its systems. The IR beam of light could be considered the robot spinal cord.

- *IR port:* Uses IR technology to transmit data to and receive data from other Palm OS handhelds, and to perform HotSync operations. Used for communication with PDA Robot's body.

- *Power button/backlight control/LED indicator:* Turns your handheld on or off and controls the backlight feature. If your handheld is turned off, pressing the power button turns the handheld on and returns you to the last screen you viewed. If your handheld is turned on, pressing the power button turns the unit off. Pressing the power button for about two seconds turns the backlight on or off. The power button also lights steadily when the handheld is charging in the cradle, and blinks to indicate alarms. Some applications enable you to set alarms to remind yourself of events or notes. You can set preferences for nonaudible alarm notification.

- *Handheld screen:* Displays the applications and information stored in your handheld. It is touch-sensitive and responds to the stylus.

4

- *Graffiti writing area:* The area where you write letters and numbers using the Graffiti alphabet.

- *Scroll buttons:* Display text and other information that extends beyond the area of the handheld screen. Pressing the lower scroll button scrolls down to view information below the viewing area, and pressing the upper scroll button scrolls up to view the information above the viewing area.

- *Application buttons:* Activate the individual handheld applications that correspond to the icons on the buttons: Date Book, Address Book, To Do List, and Note Pad. These buttons can be reassigned to activate any application on your handheld.

- *Tip:* If your handheld is turned off, pressing any application button activates the handheld and opens the corresponding application.

Beneath the Cover

PDAs are miniature versions of typical desktop systems; however, space and power consumption constraints have limited the processing power, storage space, and memory available. (This may not be true for long!) These constraints have led to very innovative designs.

Beneath the cover of each PDA is a microprocessor, which is the "brain" of the unit. All information flows in or out of it. Attached to the microprocessor are a number of peripheral devices such as the touch screen, IR port, speaker, and memory modules.

Two popular PDA microprocessors are the Intel StrongARM (**Figure 1.3**) and the Motorola DragonBall. The Intel microprocessor is typically used in devices running Windows CE, and the Motorola is used with devices running the Palm OS operating system. These processors will be described in more detail below.

ARM was established in November 1990 as Advanced RISC Machines Ltd. In 2001, more than 538 million Reduced Instruction Set Computing (RISC) microprocessors were shipped, 74.6 percent of which were based on the ARM microprocessor architecture. ARM licenses its intellectual property (IP) to a network of partners, which includes some of the world's leading semiconductor and system companies, including 19 out of the top 20 semiconductor vendors world-

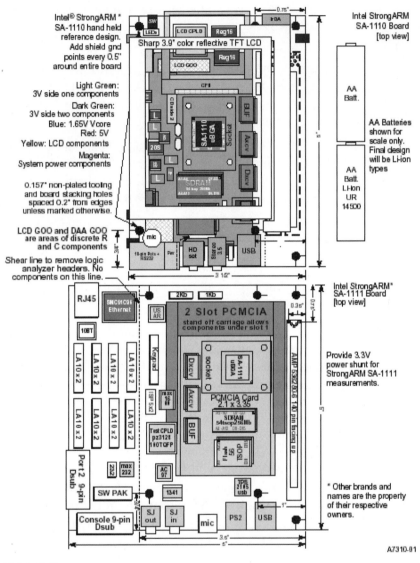

Figure 1.3

The Intel StrongARM device board SA-1110.

wide. These partners utilize ARM's low-cost, power-efficient core designs to create and manufacture microprocessors, peripherals, and system-on-chip (SoC) solutions. As the foundation of the company's global technology network, these partners have played a pivotal role in the widespread adoption of the ARM architecture. To date, ARM partners have shipped more than one billion ARM microprocessor cores!

Following is a list of ARM's key semiconductor and system partners. Obviously, this is a very well accepted architecture. 3Com, Agere, Agilent, AKM, Alcatel, Altera, AMI Semiconductor, Analog Devices, Atmel, Basis, Cirrus Logic, Cogency, Conexant, Epson, Ericsson, Fujitsu, Global UniChip, Hynix, IBM, Infineon, Intel, LinkUp Systems, LSI Logic, Kawasaki, Marvell, Micronas, Mitsubishi, Mobilan, Motorola, National Semiconductor, NEC, Oak Technology, OKI, Panasonic, Philips, Prairiecom, Qualcomm, Resonext, Rohn, Samsung, Sanyo, Sharp, Silicon Wave, SiS, Sony, ST Microelectronics, Texas Instruments, Toshiba, Triscend, Virata, Yamaha, Zarlink, and ZTEIC.

The SA-1110: An Example of ARM Architecture

The SA-1110 is a general-purpose, 32-bit RISC microprocessor with a 16 kB instruction cache (Icache), an 8 kB write-back data cache (Dcache), a minicache, a write buffer, a read buffer, an MMU, an LCD controller, and serial I/O combined in a single component. The SA-1110 provides portable applications with high-end computing performance without requiring users to sacrifice available battery time. Its power-management functionality provides further power savings. For embedded applications, the SA-1110 offers high-performance computing at consumer electronics pricing with millions of instructions per second (MIPS)-per-dollar and MIPS-per-watt advantages. The SA-1110 delivers in price/performance and power/performance, making it a choice for portable and embedded applications.

Figure 1.4 shows that the StrongARM has five serial channels used to communicate with peripheral devices. Because we will communicate primarily through the serial ports, the use for each port will be explained in detail.

- *Channel 0:* User datagram protocol (UDP) is a connectionless protocol (one in which the host can send a message without establishing a connection with the recipient) that, like transmission control protocol (TCP), runs on top of Internet protocol (IP) networks. Unlike TCP/IP, UDP/IP provides very few error recovery services, offering instead a direct way to send and receive

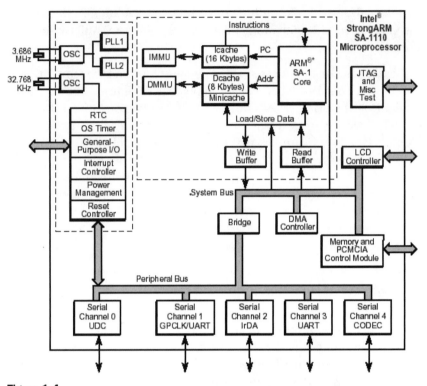

Figure 1.4

Block diagram of the Intel StrongARM SA-1110 microprocessor.

datagrams over an IP network. It is used primarily for broadcasting messages over a network. In medical imaging, UDP is used to log information from various devices to a system logging repository. A datagram is a piece of a message transmitted over a packet-switching network, and is a packet of information that contains the destination address in addition to data.

- *Channel 1:* GPCLK/UART—This channel can be used as a general purpose clock (GPCLK) or universal asynchronous receiver-transmitter (UART). See Channel 3 for a more detailed description.

- *Channel 2:* Infrared Data Association (IrDA) is a group of device manufacturers that developed a standard for transmitting data via IR light waves. Increasingly, computers and other devices (such as printers) come with IrDA ports. This enables you to transfer data from one device to another without any cables. For example, if both your laptop computer and printer have IrDA ports, you can simply put your computer in front of the printer and output a document, without needing to connect the two with a cable.

IrDA ports support roughly the same transmission rates as traditional parallel ports. The only restrictions on their use are that the two devices must be within a few feet of each other, and there must be a clear line of sight between them. The IrDA port on the PDA will be the main communication link to PDA-Bot; in essence, it will be the spinal cord. PDA Robot responds to IrDA discovery requests and identifies itself as "generic IrDA." I decided to use an IrDA data link to the Robot because it is a very reliable communication link (error correction is built into it) that requires absolutely no cables!

See: Chapter 4: Infrared Communications Overview, PDA Bot IR transponder.

- *Channel 3:* Universal asynchronous receiver-transmitter (UART): Intel provides a development board for the StrongARM SA-1100 microprocessors. It is interesting to note that most PDAs using the StrongARM are almost identical in function to that of the development board.

Increasingly, ARM-based microprocessors are being used in Palm OS devices such as the Tungsten (see **Figure 1.5**). It has a Texas Instruments OMAP1510 processor (an enhanced ARM-based processor).

The OMAP1510 processor includes the following:

- TI-enhanced ARM9 up to 175 MHz (maximum frequency).

- TMS320C55x DSP up to 200 MHz (maximum frequency).

- Voltage: 1.5v nominal.

- Optimized software architecture that allows designers to leverage dual processing, and provides a complete and seamless software foundation.

- DSP/BIOS Bridge that provides a seamless interface to the DSP using standard APIs allowing easy access to DSP multimedia algorithms.

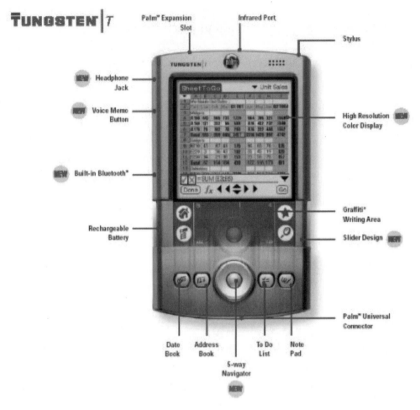

Figure 1.5

Palm OS Tungsten.

- Open platform that enables a large network of independent developers to provide a broad range of OMAP compatible software solutions.

- LCD control/frame buffer for 16-bit QVGA display.

- USB client and host control.

- MMC-SD support.

- Bluetooth interface.

- USB, uWire, camera, and enhanced audio codec interface.

- Small, 289-pin MicroStar BGA package eases design in space-constrained devices.

To provide the optimal balance of high performance and low power consumption necessary for these devices, the OMAP1510 combines the TMS320C55x DSP core with a TI-enhanced ARM925 processor.

The ARM architecture is well suited for control-type code, such as the operating system and user interface. The C55x DSP provides the additional processing power to handle the compute-intensive operations such as security, multimedia, and speech. This is a great chip for PDAs. **Figure 1.6** shows the extensively integrated OMAP microchip.

A final example of a system on a chip design is the popular MC68EZ328 (DragonBall EZ) Integrated Portable System Processor used in many of the PDAs currently in use. Even though these processors typically run at a slower clock rate, they are capable of performing 2.7 MIPS performance at 16.58 MHz processor clock, and 3.25 MIPS performance at 20 MHz processor clock—very impressive for their size and cost!

The second member of the DragonBall family, the MC68EZ328, inherits the display capability of the original DragonBall processor, but features a more flexible LCD controller with a streamlined list of peripherals placed in a smaller package. This processor is mainly targeted for portable consumer products, which require fewer peripherals and a more flexible LCD controller. By providing 3.3 V, fully static operation in efficient 100 TQFP and 144 MAPBGA packages, the MC68EZ328 delivers cost-effective performance to satisfy the extensive requirements of today's portable consumer market. A number of the Visor handspring PDAs utilize the Dragonball processors. **Figure 1.7** is the block diagram of the MC68EZ328.

Most PDAs have their small size and expandability in common, regardless of the processor or operating system. In the near future, we will likely see enough power in the palm of your hand to make the desktop computer obsolete! The prices of even the high-end PDAs have dropped dramatically over the last year, and will likely continue to do so. There are slews of very low-cost, used PDAs floating around at auctions, garage sales and in the classified ads. Even a very low-end PDA running at least Palm OS version 1.1 will be sufficient for this project. Look around if you don't have one, and you will likely find a very good deal on a used PDA.

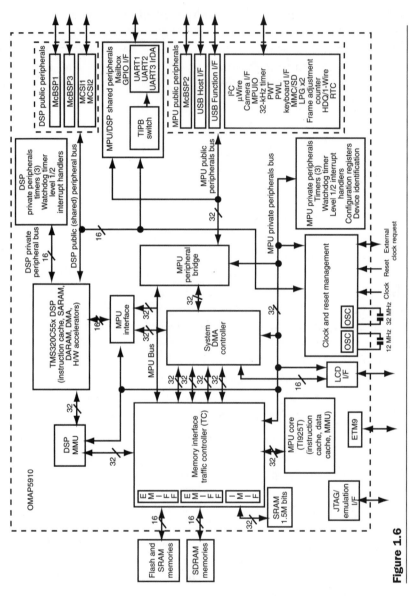

Figure 1.6

Block diagram of an OMAP processor.

12

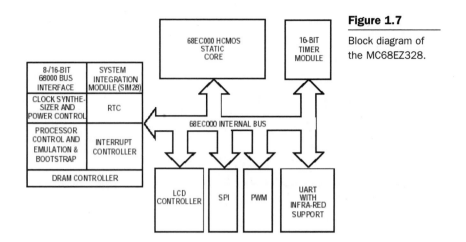

Figure 1.7

Block diagram of the MC68EZ328.

2

Robotic System Overview

PDA Robot consists of a robotic body and a PDA (handheld computer) brain. This book will guide you through the creation of PDA Robot. The project consists of mechanical, electronic, and software components. **Figure 2.1** shows PDA Robot roaming autonomously through the house, capturing images when any motion is detected. The PDA sitting on top is the machine's main controller, receiving, analyzing, and sending data to the robot body. The PDA is connected to a desktop computer that is monitoring the system, interpreting both data and the video stream. The personal computer (PC) also acts as a control station where the robot can be controlled remotely, based on the video that is displayed.

The block diagram in **Figure 2.2** is a high-level conceptualization of PDA Robot. It doesn't show the PDA connected to the wireless network.

Major Electronic Parts

Microchip MCP2150 IrDA Standard Protocol Stack Controller

The MCP2150 is a cost-effective, low pin-count (18-pin), easy to use device for implementing Infrared Data Association (IrDA) standard wireless connectivity (see **Figure 2.3**). The MCP2150 provides support for the IrDA standard protocol "stack" plus bit encoding/decoding.

Figure 2.1

PDA Robot.

Figure 2.2

Block diagram of
PDABot.

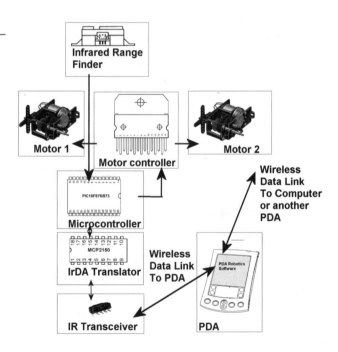

Figure 2.3

MCP 2150 chipset.

Vishay TFDS4500 Serial Infrared Transceiver

The TFDU4100, TFDS4500 (**Figure 2.4**), and TFDT4500 are a family of low-power infrared (IR) transceiver modules compliant to the IrDA standard for serial infrared (SIR) data communication, supporting IrDA speeds up to 115.2 kb/s. Integrated within the transceiver modules is a photo PIN diode, infrared emitter (IRED), and a low-power analog control integrated circuit (IC) to provide a total front-end solution in a single package. Telefunken's SIR transceivers are available in three package options, including our BabyFace package (TFDU4100), once the smallest SIR transceiver available on the market. This wide selection provides flexibility for a variety of applications and space constraints. The transceivers are capable of directly interfacing with a wide variety of I/O chips, which perform the pulse-width modulation/demodulation function, including Telefunken's TOIM4232 and TOIM3232. At a minimum, a current-limiting resistor in series with the IRED and a VCC bypass capacitor are the only external components required to implement a complete solution.

Figure 2.4

The vishay TFDS4500.

17

PIC16F876 Microcontroller

This powerful (200 nanosecond instruction execution) yet easy-to-program (only 35 single-word instructions) CMOS flash-based 8-bit microcontroller packs Microchip's powerful programmable integrated circuit (PIC) architecture into an 18-pin package, and is upwards compatible with the PIC16C7x, PIC16C62xA, PIC16C5X, and PIC12CXXX devices. The PIC16F876 features 8 MHz internal oscillator, 256 bytes of EEPROM data memory, a capture/compare/PWM, an addressable USART, and two comparators that make it ideal for advantage analog/integrated level applications in automotive, industrial, appliances, and consumer applications (see **Figure 2.5**).

PDIP, SOIC

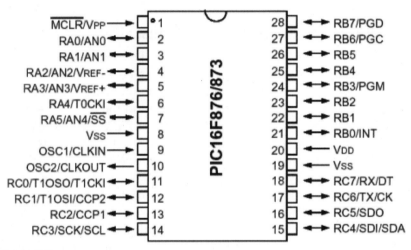

Figure 2.5

The PIC16F876.

See Chapter 7: Programming the PIC16F876 Microcontroller for more information.

L7805ACV Voltage Regulator (5 Volts)

The L7800A series of three terminal positive regulators is available in TO-220, TO-220FP, and D²PAK packages and several fixed output voltages, making it useful in a wide range of applications. These regulators can provide local on-card regulation, eliminating the distribution problem associated with single point regulation. Each type employs internal current limiting, thermal shutdown, and safe area protection, making it essentially indestructible. If adequate heat sinking is provided, they can deliver over 1A output current. Although designed primarily as fixed voltage regulators, these devices can be used with external components to obtain adjustable voltage and currents. Note: PDABot draws very little current, so heat sinking is not necessary. **Figure 2.6** shows the available packages.

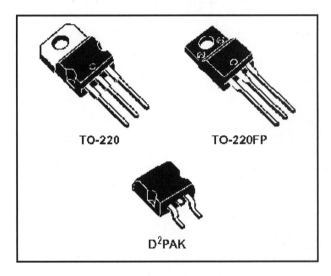

Figure 2.6

The L7800A
chipset.

TO-220   TO-220FP

D²PAK

L298 Dual Full-Bridge Driver

The L298 is used in PDA Robot to drive the two DC motors. It is an integrated monolithic circuit in 15-lead Multiwatt and Power SO20 packages. It is a high-voltage, high-current dual full-bridge driver designed to accept standard TTL logic levels and drive inductive loads such as relays, solenoids, DC, and stepping motors. Two enable inputs are provided to enable or disable the device independently of the input signals. The emitters of the lower transistors of each bridge are connected together, and the corresponding external terminal can be used for the connection of an external sensing resistor. Additional supply input is provided so that the logic works at a lower voltage. **Figure 2.7** illustrates the physical layout of the L298.

Figure 2.7

The L298 h-bridge
chipset.

Multiwatt15   **PowerSO20**

Sharp GP2D12 Infrared Range Finder

The GP2D12 is a compact, self-contained IR ranging system incorporating an IR transmitter, receiver, optics, filter, detection, and amplification circuitry (see **Figure 2.8**). Along with the wireless video camera, it gives PDA Robot a sense of sight, allowing it to navigate
autonomously around objects. The unit is highly resistant to ambient
light and nearly impervious to variations in the surface reflectivity of
the detected object. Unlike many IR systems, this has a fairly narrow
field of view, making it easier to get the range of a specific target. The
field of view changes with the distance to an object, but is no wider
than 5 cm (2.5 cm either side of center) when measuring at the maximum range.

Figure 2.8

The GP2D12.

DYN2009635 20 MH and RXDMP49 11.0952 MHz "AT" Cut Quartz Crystal Oscillator

The PIC16F876 RISC microcontroller uses a 20 MHz crystal, and the MCP2150 uses an 11 MHz crystal. While the PIC16F876 has an 8 MHz internal oscillator, a higher clock rate is desired for the communication link, analog input turnaround, and motor control reaction time via the digital outputs. **Figure 2.9** shows the physical dimensions of the crystals.

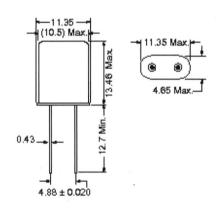

Figure 2.9

Physical dimensions of the RXDMP49 and DYN2009635 crystal oscillators. Side and bottom views.

3

Tools and Equipment

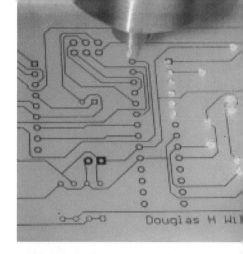

To complete the PDA Robot project, some tools like the soldering iron are essential; some simply make the job easier. The following lists the essentials and then the "nice to have equipment" you can buy when your skill in electronics and software earns you a great living, with a lot of excitement along the way!

Essential Tools and Equipment

Essentials, shown in **Figure 3.1**, include a screwdriver (A), a pair of side cutting pliers (B), a utility knife (C), a simple multimeter (D), a soldering iron (E), a ruler (F), a hack saw (G), a porcelain cooking tray, and about 45 minutes time on a drill press (www.thinkbotics.com). Buy a drill press if you plan on making a lot of circuits (see **Figure 3.2**).

Another very useful tool is a chip puller. Quite often they come with low-cost computer tool kits. When you reprogram the microchip (PIC16F876) in this project, it needs to be pulled from the board, programmed, and reinserted. You can use your hands to pull the chips, but you risk bending or squashing the pins, as well as frying chips with a jolt of static electricity. I almost put the chip puller in the essential list until the couch swallowed mine, and I was simply (carefully) pulling the chips from the board with my hand. A pair of wire cutters for clipping the leads off the electronics components is helpful, in

Figure 3.1

The essential tools.

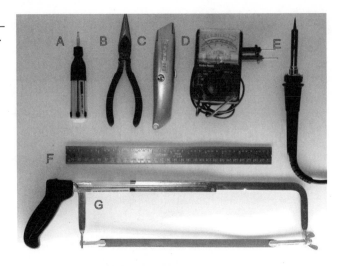

Figure 3.2

Drill press.

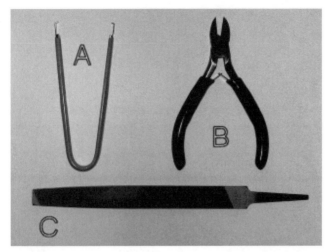

Figure 3.3

Tools.

addition to a file to smooth any metal edges. **Figure 3.3** shows a chip puller (A), wire cutters (B), and a file (C).

To make the job of soldering safe, get the tools shown in **Figure 3.4**, including a good soldering iron holder (A). When hot, it is a fire hazard. The soldering iron tip cleaner (B) makes soldering a lot faster and ensures a high-quality weld. The solder sucker (C) helps to easily remove a component or fix a bad spot.

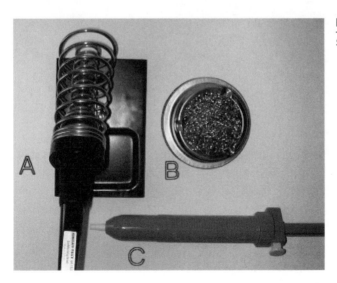

Figure 3.4

Soldering tools.

You will need four drill bits, shown in **Figure 3.5**, to complete the circuit board and body of PDA Robot. Use the 7/64 (A) to drill the holes in the aluminum plates to mount the circuits, supports, and motors. Use the 1/16 (B), 1/32 (C), and the 3/64 (D) to drill holes in the circuit for the various components.

Figure 3.5

Drill bits.

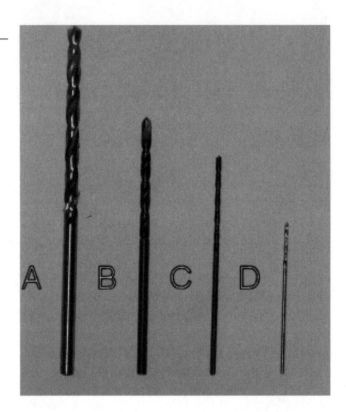

Safety First

Please do yourself a favor and buy *eye protection*. You need safety glasses when drilling and etching the circuit board. Always use common sense around any equipment. Remember to unplug your soldering iron before going out, especially if you have pets or small children.

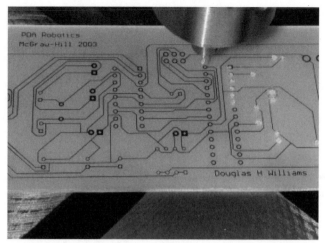

Figure 3.6

Drilling the holes on the circuit board.

Where to Get Equipment

Go to garage sales and flea markets to find some very good deals. A lot of equipment is in great shape even after collecting dust for years in people's basements. Asking for tools for birthdays and Christmas is a great way to acquire them over time if you are on a limited budget.

4

Infrared Communications Overview

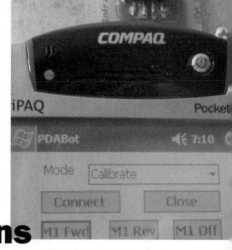

Infrared (IR) radiation lies between the visible and microwave portions of the electromagnetic spectrum, and is the medium that the personal digital assistant (PDA) uses to talk to the robot control circuitry (see **Figure 4.1**).

IR light is broken into the following three categories.

- *Near-infrared (near-IR)*—Closest to visible light, near-IR has wavelengths that range from 0.7 to 1.3 microns, or 700 billionths to 1300 billionths of a meter.

- *Mid-infrared (mid-IR)*—Mid-IR has wavelengths ranging from 1.3 to 3 microns. Both near-IR and mid-IR are used by a variety of electronic devices, including remote controls. It is in this mid range that the PDA will communicate with the robotic body using the Infrared Data Association (IrDA) communication protocol.

- *Thermal-infrared (thermal-IR)*—Occupying the largest part of the IR spectrum, thermal-IR has wavelengths ranging from 3 microns to over 30 microns.

The infrared emitters (IREDs) used for PDA devices fall into the near-IR category.

PDABot will use an IrDA protocol called IrCOMM (9-wire "cooked" service class) and the IrLMP. To simplify the task of using the IrDA

Figure 4.1

PDA Robot's IR transceiver next to an iPAQ 3850.

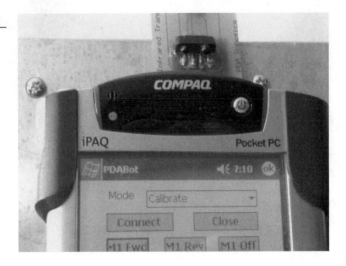

protocol, PDABot uses a Microchip MCP2150, (see **Figure 4.2**) an IrDA standard protocol stack controller, and a Vishay Telefunken TFDS4500 serial infrared transceiver (SIR 115.2 kb/s).

A widely used protocol that most devices using IR adhere to is IrDA. Both Palm OS and Windows have incarnations of IrDA, which will be explained in detail in Chapter 8: PDA Robot PalmOS Software Using Code Warrior 8 and Chapter 9: PDA Robot Software for Pocket PC 2002 (Windows CE).

IrDA is an international organization that creates and promotes inter-operable, low-cost IR data interconnection standards that support a walk-up, point-to-point user model. The Infrared Data Association

Figure 4.2

MCP2150 block diagram.

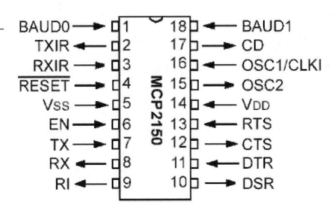

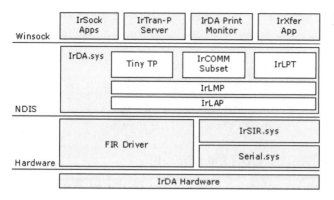

Figure 4.3

IrDA architecture.

standards support a broad range of appliances, computing, and communications devices. **Figure 4.3** illustrates Windows IrDA architecture, as defined today.

Technical Summary of IrDA Data and IrDA Control

IrDA's New Full Range of Digital Information Exchange via Cordless IR Connections

Regarding present publications on IrDA features for PC99, IrDA Data is recommended for high-speed, short-range, line-of-sight, point-to-point cordless data transfer—suitable for handheld personal computers (HPCs), PDAs, digital cameras, handheld data collection devices, etc. If IrDA is supported, it must be targeted at the 4 Mb/s components. IrDA Control is recommended for in-room cordless peripherals to hostPC. PC99 is for lower speed, full cross range, point-to-point or point-to-multipoint cordless controller—suitable for keyboards (one-way), joysticks (two-way and low latency), etc. IrDA Data and IrDA Control require designer attention to ensure spatial or time-sharing techniques, so as to avoid interference.

Since 1994, IrDA Data has defined a standard for an interoperable, universal, two-way, cordless IR light transmission data port. IrDA technology is already in over 300 million electronic devices including desktops, notebooks, palm PCs, printers, digital cameras, public phones/kiosks, cellular phones, pagers, PDAs, electronic books, electronic wallets, toys, watches, and other mobile devices.

IrDA Data protocols consist of a mandatory set of protocols and a set of optional protocols. The mandatory protocols include the following:

- Physical Signaling Layer (PHY)

- Link Access Protocol (IrLAP)

- Link Management Protocol/Information Access Service (IrLMP/ IAS)

Characteristics of Physical IrDA Data Signaling:

- Range: Continuous operation from contact to at least one (typically two can be reached). A low-power version relaxes the range objective for operation from contact through at least 20 cm between low-power devices, and 30 cm between low-power and standard-power devices. This implementation affords 10 times less power consumption. These parameters are termed the required maximum ranges by certain classes of IrDA featured devices, and set the end-user expectation for discovery, recognition, and performance.

- Bidirectional communication is the basis of all specifications.

- Data transmission from 9600 b/s with primary speed/cost steps of 115 kb/s and maximum speed up to 4 Mb/s.

- Data packets are protected using a cyclic redundancy check (CRC) (CRC-16 for speeds up to 1.152 Mb/s and CRC-32 at 4 Mb/s).

Characteristics of IrDA Link Access Protocol (IrLAP):

- Provides a device-to-device connection for the reliable, ordered transfer of data.

- Device discovery procedures.

- Handles hidden nodes.

Characteristics of IrDA Link Management Protocol (IrLMP):

- Provides multiplexing of the IrLAP layer.

- Provides multiple channels above an IrLAP connection.

- Provides protocol and service discovery via the Information Access Service (IAS).

Optional IrDA Data Protocols

The optional IrDA data protocols include the following:

- Tiny TP provides flow control on IrLMP connections with an optional segmentation and reassembly service.

- IrCOMM provides COM (serial and parallel) port emulation for legacy COM applications, printing, and modem devices.

- OBEX™ provides object exchange services similar to hypertext transfer protocol (HTTP).

- IrDA Lite provides methods of reducing the size of IrDA code, while maintaining compatibility with full implementations.

- IrTran-P provides image exchange protocol used in digital image capture devices/cameras.

- IrMC provides specifications on how mobile telephony and communication devices can exchange information. This includes phone book, calendar, and message data, as well as how call control and real-time voice are handled (RTCON) via calendar.

- IrLAN describes a protocol used to support IR wireless access to local area networks.

IrDA Control

IrDA Control is an IR communication standard that allows cordless peripherals such as keyboards, mice, game pads, joysticks, and pointing devices to interact with many types of intelligent host devices. Host devices include PCs, home appliances, game machines, and television/Web set-top boxes. IrDA Control is well suited to deal with devices that leverage the USB HID class of device controls and home appliances.

IrDA Control protocols consist of a mandatory set of protocols, including:

- PHY (Physical Layer)

- MAC (Media Access Control)

- LLC (Logical Link Control)

Characteristics of IrDA Control Physical Signaling:

- Distance and range equivalent current unidirectional IR remote control units (minimum 5 m range).

- Bidirectional communication is the basis of all specs.

- Data transmission at 75 kb/s at the top end.

- The data are coded using a 16-pulse sequence multiplied by a 1.5-MHz subcarrier, which is allocated for high-speed remote control in IEC 1603-1, although this base band scheme has harmonics that can intrude upon other IEC bands.

- Data packets are protected with a CRC (CRC-8 for short packets and CRC-16 for long packets). The physical layer is optimized for low-power usage and can be implemented with low-cost hardware.

Characteristics of IrDA Control MAC:

- Enables a host device to communicate with multiple peripheral devices (1:n) and up to eight peripherals simultaneously.

- Ensures fast response time (13.8 ms basic polling rate) and low latency.

Asymmetric MAC provides for dynamic assignment and reuse of peripheral addresses. Scheduling of media access is actually buried in the HID LLC.

Characteristics of the IrDA Control LLC:

- Provides reliability features that provide data sequencing and retransmission when errors are detected.

- Works with an HID-IrDA control bridge to enable the link control functions of USB-HID.

- All required and optional layers of the IrDA Data and IrDA Control specifications are described in specifications that can be downloaded at no charge from the IrDA Web site: www.irda.org. Interop product registration is strongly advised on this site.

IrDA specifications are now supported by all divisions of Microsoft (IDG, WinCE, Win98, Win2000, and Windows XP), and this universal data port is recommended on PC99 products (mandated on certain WinCE products—PalmPC, etc.)

PDA Robot will use the IrDA Data protocol, not the IrDA Control protocol, to ensure a reliable high-speed bidirectional flow of data between the body and the brain (PDA). All decisions will be made on-board the PDA, using the software outlined in this book.

Windows CE (Pocket PC) and IrDA

One of the key features of Windows CE-based devices is the ability to communicate with other devices. Windows CE supports two basic types of communication: serial communication and communication over a network. Most devices feature built-in communications hardware, such as a serial port or an IR transceiver. The network driver interface specification (NDIS) implementation on Windows CE supports the following communications media: Ethernet (802.3), Token Ring (802.5), IrDA, and wide area network (WAN). The diagram shown in **Figure 4.4** outlines the communications architecture of the Windows CE operating system, specifically the components of the IrDA protocol layer and how IrDA miniport drivers communicate through the NDIS library, with their network interface cards (NICs) and applications.

In the Windows CE communications architecture, the NDIS interface is located below the IrDA, transmission control protocol/Internet protocol (TCP/IP), and point-to-point protocol (PPP) drivers. The NDIS wrapper presents an interface to the upper and lower edges of a miniport driver. To an upper-level driver, such as the TCP/IP protocol driver, the NDIS interface looks like a miniport driver. To the miniport, the NDIS interface looks like an upper-level protocol driver. On the bottom of the communications architecture, the NDIS interface functions as a network adapter driver that interfaces directly with the network adapter at the lower edge. At the upper edge, the network adapter driver presents an interface to allow upper layers to send packets on the network, handle interrupts, reset or halt the network adapter, and query or set the operational characteristics of the driver.

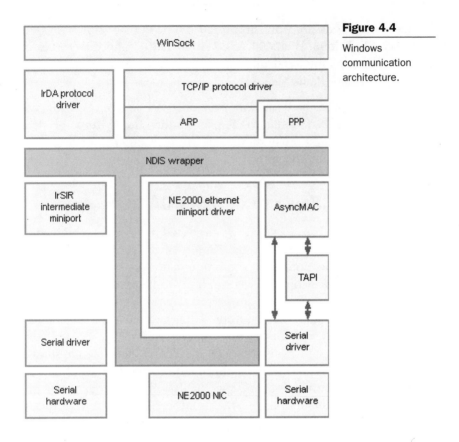

Figure 4.4

Windows communication architecture.

Communication Link Speeds

Unlike typical NDIS media, the IR medium supports a large number of different speeds for transmitting and receiving bits. Current definitions for operating speed vary from 2400 bits per second (b/s) to 16 megabits per second (Mb/s). In the future, more speeds may be defined by IrDA. Varying design goals at different speeds have led to different coding methods for frames: SIR, MIR, FIR, and VFIR. The differences in frame coding methods must be handled by the IrDA miniport driver and be transparent to the protocol.

The currently defined IrDA speeds and their corresponding frame coding methods (Serial IrDA [SIR] link speeds, Medium IrDA [MIR] link speeds, Fast IrDA [FIR] link speeds, and Very Fast IrDA [VFIR] link speeds) are listed in **Table 4.1**.

Speed (in bps)	Frame Coding Method	Table 4.1
2400	SIR	IrDA Speeds and
9600	SIR	Corresponding
19,200	SIR	Frame Coding
38,400	SIR	Methods
57,600	SIR	
115,200	SIR	
576,000	MIR	
1.152 Mb/s	MIR	
4 Mb/s	FIR	
16 Mb/s	VFIR	

Communication Link Turnaround Times

An IR adapter consists of an IR transceiver, along with supporting hardware for encoding and decoding frames. This IR transceiver contains a transmitter light-emitting diode (LED) and a receiver diode that are typically located quite close together. The receiver diode is sensitive to IR light because it must receive transmissions from a remote IR LED over distances up to at least 1 m. The transmitter LED is quite powerful because it must transmit to a remote receiver diode over the same distances.

During transmission, a local LED typically emits enough light to saturate the local receiver diode. In much the same way that it is difficult for people to see well after staring at the sun, it is difficult for the local receiver diode to correctly receive incoming frames immediately after the local LED transmits outgoing frames.

To allow time for the local receiver diode to recover from the saturation state and become capable of again receiving incoming frames, the IrDA protocol defines a parameter known as turnaround time. Turnaround time specifies the amount of time, in milliseconds, that it takes the receiver diode to recover from saturation. In some IrDA devices, the turnaround time may be negligible; in other IrDA devices, it can be a relatively long period of time.

The turnaround time of the local receiver diode does not affect the behavior of the local transceiver. However, the turnaround time of the local receiver diode affects the anticipated behavior of the remote transceiver. For example, if a local transceiver requires a 1-ms delay from the time its LED finishes transmitting to the time its receiver diode is capable of receiving, the remote station must wait 1 ms from receiving the last bit of a frame before beginning to transmit a new frame. The remote station performs this wait to honor the local transceiver's turnaround time.

To honor the turnaround time of the remote transceiver, the IrLAP protocol might sometimes specify to delay transmission of a packet. To do so, the IrLAP protocol specifies the amount of time before a packet should be transmitted. The IrDA miniport driver must not transmit the packet before waiting the requested amount of time, although the driver can wait longer if necessary. The IrLAP protocol specifies transmission delay time of a packet in the media-specific member of the packet's associated out-of-band (OOB) data block.

IrLAP defines the format of the frames sent and received on the IR media. Each IrLAP frame consists of the following elements:

- One or more beginning of frame (BOF) flags that mark the beginning of the frame. The size of the BOF member varies in length, depending on the speed.

- An address (A) member that identifies the secondary connection address. The address member is 8 b. The address member specifies the address of a device that belongs to a particular IrDA miniport driver. This IrDA miniport driver transmits or receives the frame that contains this address through this device.

- A control (C) member that specifies the function of the particular frame. The control member is 8 b.

- An optional information (I) member that contains the information data. The information member is an integral number of octets.

- A frame check sequence (FCS) member that allows the receiving station to check the transmission accuracy of the frame. The FCS member is either 16 or 32 b, depending on the speed.

- An end of frame (EOF) flag that signals the end of the frame. The size of the EOF member varies, depending on the speed.

The following example of an IrLAP frame shows the order of the elements that were described in the preceding section.

BOF	A	C	I		FCS	EOF

SIR Coding

This topic describes how IrDA miniport drivers or their IR NICs code frames for transmission at Serial IrDA (SIR) link speeds. The SIR specification defines a short-range IR asynchronous serial transmission mode with one start bit, eight data bits, and one stop bit. The maximum data rate is 115.2 Kb/s (half duplex). This SIR coding scheme is called return to zero, inverted (RZI). The primary benefit of coding frames for SIR speeds is that existing serial hardware can be used very cheaply. The low cost of using serial hardware is one of the reasons for the widespread availability of IR SIR devices.

The BOF flag for SIR speeds is defined as 0xC0. The EOF value is defined as 0xC1. To avoid ambiguity in a frame that contains BOF and EOF, an escape sequence is defined for values of 0xC0 and 0xC1 that occur in other parts of the frame. The escape character is defined as 0x7D.

For each byte that the transmitter encounters that is the same as a BOF, EOF, or the escape character, the transmitter performs the following steps:

1. Inserts a control-escape byte (0x7D) preceding such a byte.

2. Complements bit five of each byte that is the same as the BOF, EOF, or escape character (i.e., performs an exclusive OR operation on such a byte with 0x20).

MIR Coding

This topic describes how IrDA miniport drivers and their IR NICs code frames for transmission at Medium IrDA (MIR) link speeds. The MIR data rates are 0.576 Mb/s and 1.152 Mb/s (half duplex).

For MIR link speeds, definitions for BOF and EOF values are the same; both BOF and EOF are defined as 0x7E. To avoid ambiguity in the frame with BOF and EOF, rather than using an escape sequence as is done at SIR rates, a zero is inserted at MIR rates after any five consecutive one bits in all members, except BOF and EOF. Because the process of inserting and stripping zeros at the bit level is highly processor-intensive, it is strongly recommended that this logic be implemented in hardware. At MIR link speeds, two BOF flags are required on every frame.

For MIR link speeds, the CRC used is the same as for SIR speeds. That is, for MIR link speeds, the IrDA miniport driver also typically calculates the CRC value, rather than the driver's hardware.

FIR Coding

This topic describes how IrDA miniport drivers and their IR NICs code frames for transmission at Fast IrDA (FIR) link speeds. The FIR specification defines short-range, low-power operation at 4 Mb/s (half duplex). All FIR devices are also required to support SIR operation.

For FIR link speeds, an entirely different coding scheme, called four pulse position modulation (4PPM), is used. The 4PPM coding scheme defines special flags for BOF and EOF. Always implement the 4PPM coding scheme in hardware.

The IrDA miniport driver may still be required to calculate the CRC to validate the frame. For FIR link speeds, a 32-bit CRC is used. An algorithm for calculating the 32-bit CRC is available in the publication *Infrared Data Association Serial Infrared Physical Layer Link Specification*, available from IrDA.

VFIR Coding

This topic describes how IrDA miniport drivers and their IR NICs code frames for transmission at Very Fast IrDA (VFIR) link speeds. The VFIR specification defines short-range, low-power operation at 16 Mb/s (half duplex). All VFIR devices are also required to support FIR and SIR operation.

For VFIR link speeds, an entirely different coding scheme, called HHH(1,13), is used. The letters HHH that represent this coding scheme are the initials of the three researchers who invented it. Always implement the HHH(1,13) coding scheme in hardware. For more information on HHH(1,13), see the publication *Infrared Data Association Serial Infrared Physical Layer Link Specification*, available from IrDA.

The IrDA miniport driver's hardware can calculate the CRC to validate the frame. However, if hardware does not calculate CRC, the IrDA miniport driver must calculate CRC. For VFIR link speeds, a 32-bit CRC is used, which is the same as that used for FIR link speeds. An algorithm for calculating the 32-bit CRC is available in the publication *Infrared Data Association Serial Infrared Physical Layer Link Specification.*

The IrDA specification will give you an idea of the technical details involved in the protocol. When we write to the software, you will find it is not as complicated as it seems. The creators of the Windows and Palm operating systems gave an application programming interface (API) that makes creating an association, sending, and receiving data a fairly straightforward task.

5

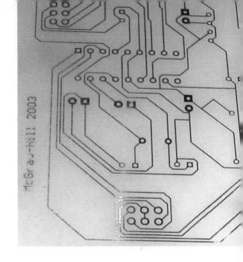

The
Electronics

This chapter consists of two parts. First is an overview of the electronic design, focusing on various portions of the schematic diagram. Second is a description of each component, including its function and how it interacts with the others. The next chapter will explain step-by-step how to create the circuit, from "burring the board" to soldering each component.

System Overview

The circuit consists of three parts that can be separated, as I have with this project, or kept together. The main board is connected to the infrared (IR) transceiver and the motor controller circuit via 6-wire ribbon connectors. I chose to do this so that the motor circuit and the transponder could be placed anywhere, allowing for flexibility of design. The artwork for the circuit in **Figure 5.1** shows the three components of the circuit.

Figure 5.2 shows the topside of the boards (with the top personal digital assistant (PDA) support plate removed) and how they have been positioned on PDA Robot.

Not all PDAs have the IR port in the same position, so the ribbon connector lets you position the PDA in any direction. You can easily cut

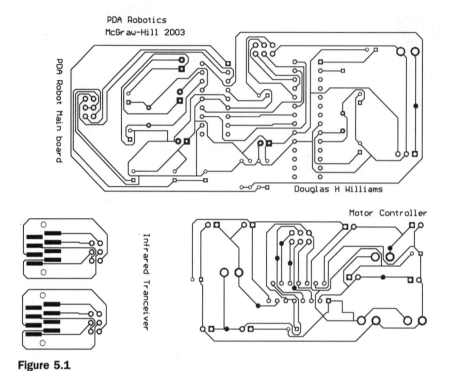

Figure 5.1

The circuit layout: Main board, motor controller, and the infrared transceiver (only one is needed).

Figure 5.2

The main board (A), infrared transceiver (B), and the motor controller (C) mounted to the bottom plate.

Figure 5.3

IPAQ and Visor
PDAs.

a slot on the top plate and stand the PDA vertically. **Figure 5.3** shows an iPAQ and a Visor positioned next to the transceiver.

For this project, I used the MG Chemical process to create the circuit board. Protel 98 SE was used to create the schematic diagrams and printed circuit board (PCB) artwork used in the MG chemical process. **Figure 5.4** shows the main portion of the circuit board after it was

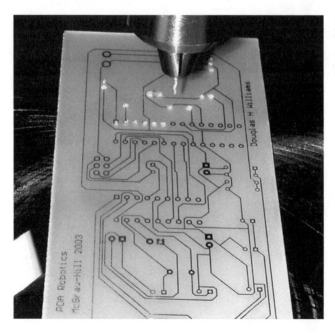

Figure 5.4

The main circuit
after exposure and
etching.

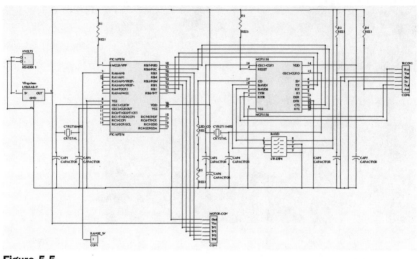

Figure 5.5

Schematic diagram of the main circuit board.

exposed and etched. It is being drilled in preparation for placement of the components. **Figure 5.5** shows the schematic diagram of the main circuit board.

Setting the Baud Rate

The MCP2150 baud rate lines, pins 1 and 18, are connected to the 8-pin duel in-line packet (DIP) switch. Pins 5 and 7 are connected to ground (low), and pins 6 and 8 are high (+5 V). This allows us to set the baud rate that the MCP2150 communicates with PIC169876. It is interesting to note that the baud rate at which the MCP2150 communicates with the PDA through the IR transceiver is independent of this setting (see **Figure 5.6**). The actual IR baud rate is determined during the handshake phase of the Infrared Data Association (IrDA) negotiation and is transparent to users. The only parameter users can set is the maximum baud rate that can be negotiated (this is explained in the software chapters for the Palm OS and Windows handhelds).

Table 5.1 shows the DIP switch settings and the associated baud rates.

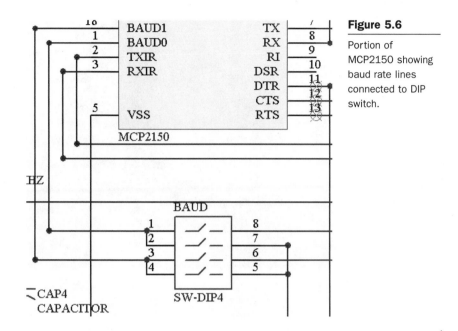

Figure 5.6

Portion of MCP2150 showing baud rate lines connected to DIP switch.

Table 5.1

DIP Switch Settings and Associated Baud Rates

DIP Switch 1 (Baud 0)	DIP Switch 2 (Baud 0)	DIP Switch 3 (Baud 1)	DIP Switch 4 (Baud 1)	Baud Rate at 11.0592 MHz
Off	On	Off	On	9600
On	Off	Off	On	19200
Off	On	On	Off	57600
On	Off	On	Off	115200

Figure 5.7 shows a close-up of the DIP switches with the baud of the MCP2150 set to 115200.

The MCP2150 Connection to the IR Transceiver

The IR transponder used by PDA Robot consists of a Microchip MCP2150 IrDA standard protocol stack controller and a Vishay TFDS4500 serial IR transceiver (a TFDU6102 fast IR transceiver can be used as well). The transceiver contains the IR emitter and receiver and the MCP2150 handles the IrDA handshaking and data exchange between the Robot and the PDA.

47

Figure 5.7

PDA Robot with the
baud rate set to
115200.

The components required for the Vishay TFDS4500 transceiver are located on the main board circuit, and the actual transceiver itself is connected via the ribbon cable. **Figure 5.8** shows the schematic diagram connection of the MCP2150's IR output (pin 2: TXIR) and input pins (pin 3: RXIR) connected to the ribbon cable connector.

The IR transceiver schematic shows the pins of the transceiver tied to the appropriate connector pins that line up with those on the main board. **Figures 5.9** and **5.10** illustrate this.

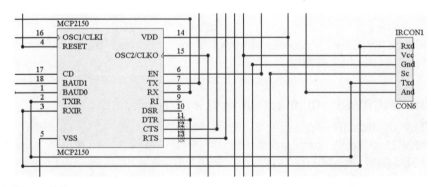

Figure 5.8

Schematic diagram showing MCP2150 connections to the IR transceiver.

TFDS4500

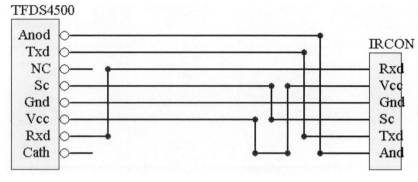

Figure 5.9

IR transceiver schematic.

The MCP2150 Connection to the PIC16F876 Microcontroller

The microcontroller is connected to the MCP2150 IrDA protocol stack decoder via the microcontroller's configurable B port. The block diagram in **Figure 5.11** shows the relationship between the transceiver, the controller, and the MCP2150.

The schematic diagram in **Figure 5.12** shows the actual pin connections between the PIC16F876 and the MCP2150. RBO is configured as

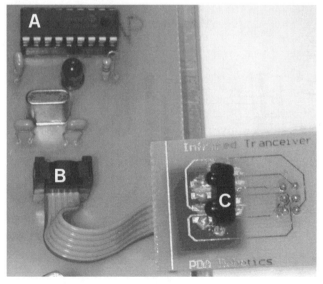

Figure 5.10

MCP2150 (A), the ribbon connection (B), and the transceiver (C).

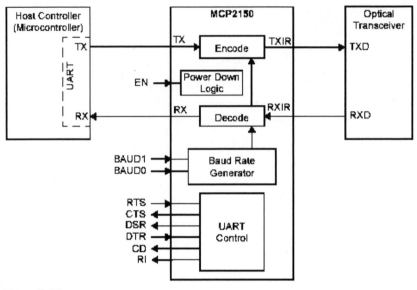

Figure 5.11

The PIC16F876, MCP2150, and TFDS4500 block diagram.

the RS232 transmit pin and RB1 as the receive pin. Pins RB6 and RB7 are configured as inputs, used to monitor the MCP2150's Request to send (RTS: pin 13) and Clear to send (CTS: pin 12) pins. RB2, RB3, RB4, and RB5 are configured as digital outputs used to switch the L298 motor controller.

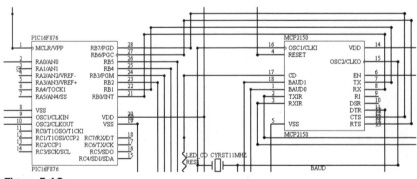

Figure 5.12

Schematic of PIC16F876 connection to MCP2150.

The Motor Controller Circuit

The motor controller circuit is connected to the main board through a six-wire ribbon connector, and has an independent load (separate from the logic) used to power the motors and the IR range finder. The power and ground for the L298's logic is carried from the main board through the ribbon cable, along with the data lines. The L298 requires that the grounds for the logic and the load must be common, so powering the logic from the regulated supply of the main board works out well. **Figure 5.13** shows the schematic diagram of the motor controller portion of the circuit. **Figure 5.14** shows the physical layout of the motor controller and the ribbon cable that connects it to the main board.

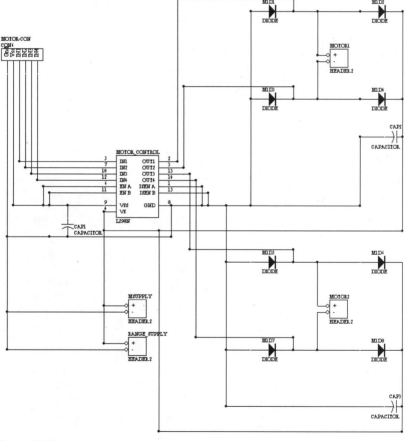

Figure 5.13

Motor controller schematic diagram.

Figure 5.14

Motor controller
PCB motor 1
connector (A),
ribbon connector
(B), motor 2
connector (C),
motor power supply
connector (D), range
finder power
connection (E),
L298 motor
controller chip (F),
and diode (G).

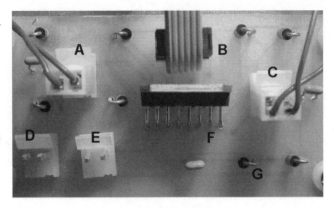

The Sharp GPD12 IR Range Finder

The Sharp GPD12 IR range finder is connected to the first configurable analog pin on the PIC16F876. **Figure 5.15** shows the pin (C), which is connected to the analog output of the range finder and the analog input of the microchip.

Figure 5.15

Left side of main
circuit, (A) 9 V
power connector,
(B) resistor, (C)
range finder input
pin, (D) capacitor,
(E) 20.0000 MHz
crystal oscillator, (F)
+5 voltage
regulator, (G)
16F876 micro-
controller, (H) and
motor controller
ribbon connector.

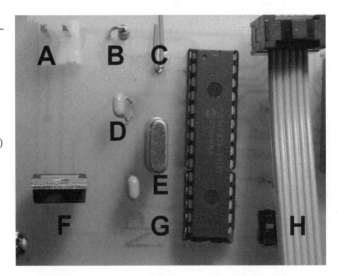

Component Descriptions

The Vishay TFDS4500

The TFDU4100, TFDS4500, and TFDT4500 are a family of low-power IR transceiver modules compliant to the IrDA standard for serial infrared (SIR) data communication, supporting IrDA speeds up to 115.2 kb/s. Integrated within the transceiver modules is a photo PIN diode, infrared emitter (IRED), and a low-power analog control integrated circuit (IC) to provide a total front-end solution in a single package. Telefunken's SIR transceivers are available in three package options, including our Baby Face package (TFDU4100), once the smallest SIR transceiver available on the market. This wide selection provides flexibility for a variety of applications and space constraints. The transceivers are capable of directly interfacing with a wide variety of I/O chips, which perform the pulse-width modulation/demodulation function, including Telefunken's TOIM4232 and TOIM3232. At a minimum, a current-limiting resistor in series with the IR and a VCC bypass capacitor are the only external components required to implement a complete solution, as is the case with PDA Robot.

TFDS4500 Features:

- Compliant to the latest IrDA physical layer standard (up to 115.2 kb/s).

- 2.7 to 5.5 V wide operating voltage range.

- Low power consumption (1.3 mA supply current).

- Power sleep mode through VCC1/SD pin (5 nA sleep current).

- Long range (up to 3.0 m at 115.2 kb/s).

- Three surface-mount package options—universal (9.7 × 4.7 × 4.0 mm), side view (13.0 × 5.95 × 5.3 mm), top view (13.0 × 7.6 × 5.95 mm).

- Directly interfaces with various super I/O and controller devices.

- Built-in electromagnetic interference (EMI) protection—no external shielding necessary.

- Few external components required.

- Backward compatible to all Telefunken SIR IR transceivers.

Package Options

TFDU4100
Baby Face (Universal)

TFDS4500
Side View

TFDT4500
Top View

Figure 5.16

Transceiver package options.

Figure 5.16 shows the three packages available. PDA Robot is using the side mount package (TFDS).

The transceiver conveniently contains an amplifier, comparator, drivers, ACG logic, the IRED, and receiver, as seen **Figure 7.17**.

Figure 5.18 shows the recommended circuit to use with the transceiver. The outlined components described as optional have been included in the design of PDA Robot. The capacitor is used to clean up any noise normally caused by the power supply. The noise being suppressed comes mostly from the two DC motors used in this project. The capacitors on the motor control circuit and those tied to the MCP2150 and TFDS4500 are used for logic circuit noise suppression.

The only required components for designing an IrDA 1.2 compatible design using Telefunken SIR transceivers are a current limiting resis-

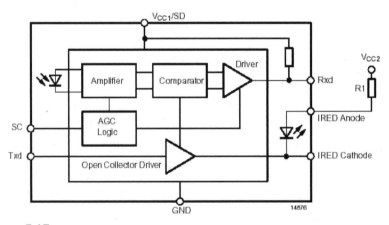

Figure 5.17

Transceiver block diagram.

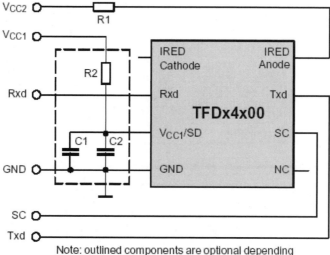

Figure 5.18

Recommended circuit diagram.

tor to the IRED. However, depending on the entire system design and board layout, additional components may be required. It is recommended that the capacitors C1 and C2 be positioned as near as possible to the transceiver power supply pins. A tantalum capacitor should be used for C1, while a ceramic capacitor should be used for C2 to suppress radio frequency (RF) noise. Also, when connecting the described circuit to the power supply, use low impedance wiring.

R1 is used for controlling the current through the IRED. To increase the output power of the IRED, reduce the value of the resistor. Similarly, to reduce the output power of the IRED, increase the value of the resistor. For typical values of R1, see **Figure 5.19**. For example, for IrDA-compliant operation (VCC2 = 5 V ± 5%), a current control resistor of 14 ohms is recommended. The upper drive current limitation is dependent on the duty cycle, and is given by the absolute maximum ratings on the data sheet and the eye safety limitations given by IEC825–1. R2, C1 and C2 are optional and dependent on the quality of the supply voltage VCC1 and injected noise. An unstable power supply with dropping voltage during transmission may reduce sensitivity (and transmission range) of the transceiver.

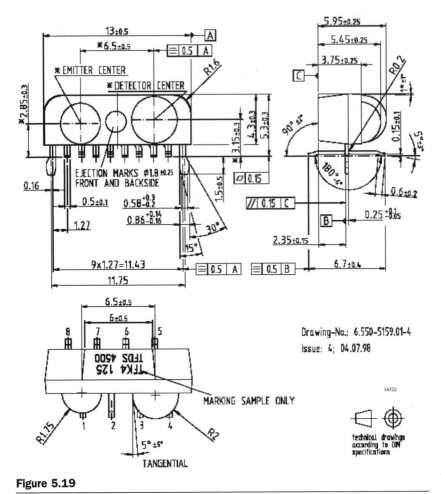

Figure 5.19

Physical dimensions of the side view package used in PDA Robot.

The sensitivity control (SC) pin allows the minimum detection irradiance threshold of the transceiver to be lowered when set to a logic HIGH. Lowering the irradiance threshold increases the sensitivity to IR signals and increases transmission range up to 3 m. However, setting the Pin SC to logic HIGH also makes the transceiver more susceptible to transmission errors, due to an increased sensitivity to fluorescent light disturbances.

It is recommended that the pin SC be set to logic LOW or left open, if the increased range is not required or if the system will be operating in bright ambient light.

This SC pin has been driven LOW in the PDA Robot circuit. However, if you decide to modify the circuit, I recommend putting a switch on the board or tying this line to a pin on the microcontroller. This would allow you to set SC high or low physically through the switch or programmatically through the microcontroller. This would enable you to hold the PDA at a much further distance from the craft when testing and calibrating the system. You could also use the PDA as a remote control.

The guide pins on the side-view and top-view packages are internally connected to ground, but should not be connected to the system ground, to avoid ground loops. They should be used for mechanical purposes only and should be left floating.

PDA Robot does not ground the guide pins. They are used only to help secure the unit to the PCB.

Shutdown.   The internal switch for the IRED in Telefunken SIR transceivers is designed to be operated like an open collector driver. Thus, the VCC2 source can be an unregulated power supply, while only a well-regulated power source with a supply current of 1.3 mA connected to VCC1/SD is needed to provide power to the remainder of the transceiver circuitry in receive mode. In transmit mode, this current is slightly higher (approximately 4 mA average at 3 V supply current), and the voltage is not required to be kept as stable as in receive mode. A voltage drop of VCC1 is acceptable down to about 2.0 V when buffering the voltage directly from the pin VCC1 to GND; see **Figure 5.20a.** This configuration minimizes the influence of high-current surges from the IRED on the internal analog control circuitry of the transceiver and the application circuit. Also, board space and cost savings can be achieved by eliminating the additional linear regulator normally needed for the IRED's high current requirements.

The transceiver can be very efficiently shut down by keeping the IRED connected to the power supply VCC2, but switching off VCC1/SD. The power source to VCC1/SD can be provided directly from a microcontroller. In shutdown, current loss is realized only as leakage current through the current-limiting resistor to the IRED (typically 5 nA). The settling time after switching VCC1/SD on again is approximately 50 μs. Telefunken's TOIM3232 interface circuit is designed for this shutdown feature. The VCC_SD, S0, or S1 outputs on the TOIM3232 can be used to power the transceiver with the necessary supply current. If the microcontroller or the microprocessor is unable to drive the supply current required by the transceiver, a low-cost SOT23 pnp transistor can be used to switch voltage on and off from the regulated power supply. The additional component cost is minimal, and saves the system designer additional power supply costs.

The 5-V regulator on the main board powers the transceiver in PDA Robot.

The Microchip MCP2150 Plug and Play IrDA

The MCP2150 is a cost-effective, low pin-count (18-pin), easy to use device for implementing IrDA standard wireless connectivity. The MCP2150 provides support for the IrDA standard protocol "stack," plus bit encoding/decoding.

The serial interface baud rates are user selectable to one of four IrDA standard baud rates between 9600 baud and 115.2 kbaud (9600, 19200, 57600, 115200). The IR baud rates are user selectable to one of five IrDA standard baud rates between 9600 baud and 115.2 kbaud (9600, 19200, 37400, 57600, 115200). The serial interface baud rate will be specified by the BAUD1:BAUD0 pins, while the IR baud rate is specified by the primary device (during Discover phase). This means that the baud rates do not need to be the same.

The MCP2150 operates in data terminal equipment (DTE) applications and sits between a UART and an IR optical transceiver. The MCP2150 encodes an asynchronous serial data stream, converting each data bit to the corresponding IR formatted pulse. IR pulses received are decoded and then handled by the protocol handler state machine. The protocol handler sends the appropriate data bytes to the host controller in UART formatted serial data.

The MCP2150 supports point-to-point applications, that is, one primary device and one secondary device. The MCP2150 operates as a secondary device. It does not support multipoint applications. Sending data using IR light requires some hardware and the use of specialized communication protocols. These protocol and hardware requirements are described, in detail, by the IrDA standard specifications.

The chapters dealing with the software for the PDAs explain, in detail, how to implement the specialized communication protocols.

The encoding/decoding functionality of the MCP2150 is designed to be compatible with the physical layer component of the IrDA standard. This part of the standard is often referred to as "IrPHY." The complete IrDA standard specifications are available for download from the IrDA Web site (www.IrDA.org).

MCP2150 Applications: PDA Robot

The MCP2150 infrared communications controller supporting the IrDA standard provides embedded system designers the easiest way to implement IrDA standard wireless connectivity. **Figure 5.20a** shows a typical application block diagram. IR communication is a wireless two-way data connection, using IR light generated by low-cost transceiver signaling technology. This provides reliable communication between two devices. Reliability is the main reason I chose this protocol and this chip. It certainly simplifies the task of creating the PDA-to-Robot data link. You can port the PDA code to the PIC microcontroller if you have the time.

IR technology has the following advantages:

- Universal standard for connecting portable computing devices.

- Easy, effortless implementation.

- Economical alternative to other connectivity solutions.

- Reliable, high-speed connection.

- Safe to use in any environment (can even be used during air travel).

- Eliminates the hassle of cables and the possibility of damage to your PDA.

- Allows PCs and other electronic devices (such as PDAs, cell phones, etc.) to communicate with each other. In this case it allows the PDA to communicate with PDA Robot.

- Enhances mobility by allowing users to easily connect.

The MCP2150 allows the easy addition of IrDA standard wireless connectivity to any embedded application that uses serial data. **Figure 5.20a** shows typical implementation of the MCP2150 in an embedded system.

Figure 5.20a

A typical application block diagram.

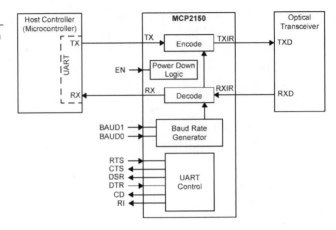

Table 5.2 describes the MCP2150 pins for the 18-pin dual in-line package used in PDA Robot's circuit.

Table 5.2

MCP2150 DIP Pin Descriptions

Pin Name	Pin #	Pin Type	Buffer Type	Description
BAUD0	1	I	ST	BAUD1:BAUD0 specify the baud rate of the device.
TXIR	2	O	–	Asynchronous transmit to IR transceiver.
RXIR	3	I	ST	Asynchronous receive from IR transceiver.
RESET	4	I	ST	Resets the device.
VSS	5	–	P	Ground reference for logic and I/O pins.
EN	6	I	TTL	Device enable. 1 = Device is enabled. 0 = Device is disabled (low power). MCP2150 only monitors this pin when in the NDM state.
TX	7	I	TTL	Asynchronous receive; from host controller UART.
RX	8	O	–	Asynchronous transmit; to host controller UART.
RI	9	–	–	Ring indicator. The value on this pin is driven high.
DSR	10	O	–	Data Set Ready. Indicates that the MCP2150 has completed reset. 1 = MCP2150 is initialized. 0 = MCP2150 is not initialized.

(continued on next page)

Table 5.2

MCP2150 DIP Pin Descriptions (continued)

Pin Name	Pin #	Pin Type	Buffer Type	Description
DTR	11	I	TTL	Data Terminal Ready. The value of this pin is ignored once the MCP2150 is initialized. It is recommended that this pin be connected so that the voltage level is either VSS or VCC. At device power up, this signal is used with the RTS signal to enter device ID programming. 1 = Enter Device ID programming mode (if RTS is cleared). 0 = Do not enter Device ID programming mode.
CTS	12	O	_	Clear to Send. Indicates that the MCP2150 is ready to receive data from the host controller. 1 = Host controller should not send data. 0 = Host controller may send data.
RTS	13	I	TTL	Request to Send. Indicates that a host controller is ready to receive data from the MCP2150. The MCP2150 prepares to send data, if available. 1 = Host controller not ready to receive data. 0 = Host controller ready to receive data. At device power up, this signal is used with the DTR signal to enter device ID programming. 1 = Do not enter device ID programming mode. 0 = Enter device ID programming mode (if DTR is set).
VDD	14	_	P	Positive supply for logic and I/O pins.
OSC2	15	O	_	Oscillator crystal output.
OSC1/CLKIN	16	I		CMOS Oscillator crystal input/external clock source input.
CD	17	O	_	Carrier Detect. Indicates that the MCP2150 has established a valid link with a primary device. 1 = An IR link has not been established (No IR Link). 0 = An IR link has been established (IR Link).
BAUD1	18	I	ST	BAUD1:BAUD0 specify the baud rate of the device.

Legend: TTL = TTL compatible input; I = Input; P = Power; ST = Schmitt Trigger input with CMOS levels; O = Output; CMOS = CMOS compatible input

Power Up. **Figure 5.20b** shows the pin's physical layout conforming to the numbering convention of first pin to the top left and the numbers wrapping around the bottom of the chip so that pin 1 is opposite pin 18.

Any time the device is powered up (parameter D003), the Power Up Timer delay (parameter 33) occurs, followed by an Oscillator Start-up

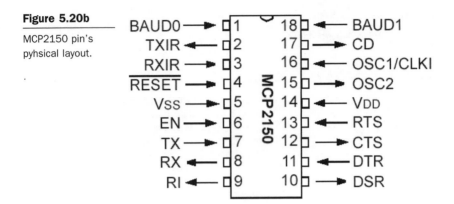

Figure 5.20b

MCP2150 pin's pyhsical layout.

Timer (OST) delay (parameter 32). Once these delays complete, communication with the device may be initiated. This communication is from both the IR transceiver's side, as well as the controller's UART interface.

Device Reset. The MCP2150 is forced into the reset state when the RESET pin is in the low state. Once the RESET pin is brought to a high state, the Device Reset sequence occurs. Once the sequence completes, functional operation begins.

Clock Source. The MCP2150 requires a clock source to operate. The frequency of this clock is 11.0592 MHz (electrical specification parameter 1A). This clock can be supplied by either a crystal/resonator or as an external clock input.

Crystal Oscillator/Ceramic Resonators

A crystal or ceramic resonator can be connected to the OSC1 and OSC2 pins to establish oscillation (**Figure 5.21**). The MCP2150 oscillator design requires the use of a parallel cut crystal. Use of a series cut crystal may give a frequency outside of the crystal manufacturer's specifications.

PDA Robot uses 22 pf capacitors for both the MCP2150 and PIC16F876. The values can range from 10 to 22 pf for a ceramic resonator and 15 to 30 pf for a crystal oscillator. Because PDA Robot uses

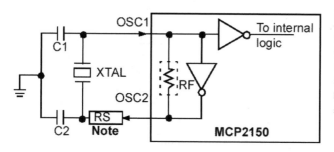

Figure 5.21

Crystal operation (or ceramic resonator). Note: A series resistor may be required for AT strip cut crystals.

crystal oscillators, the 22 pf value provides good stability and an average start-up time. It also allows us to simply swap in a ceramic resonator if desired.

Higher capacitance increases the stability of the oscillator, but also increases the start-up time. The resistor (RS) may be required to avoid overdriving crystals with low drive level specification. Since each crystal has its own characteristics, the user should consult the crystal manufacturer for appropriate values of external components.

Bit Clock

The device crystal is used to derive the communication bit clock (BIT-CLK). There are 16 BITCLKs for each bit time. The BITCLKs are used for the generation of the start bit and the eight data bits. The stop bit uses the BITCLK when the data are transmitted (not for reception). This clock is a fixed frequency and has minimal variation in frequency (specified by crystal manufacturer).

UART Interface

The UART interface communicates with the controller. This interface is a half-duplex interface, meaning that the system is either transmitting or receiving, but not both simultaneously.

Baud Rate

The baud rate for the MCP2150 serial port (the TX and RX pins) is configured by the state of the BAUD1 and BAUD0 pins. These two device pins are used to select the baud rate at which the MCP2150 will transmit and receive serial data (not IR data).

Transmitting

When the controller sends serial data to the MCP2150, the controller's baud rate is required to match the baud rate of the MCP2150's serial port.

Receiving

When the controller receives serial data from the MCP2150, the controller's baud rate is required to match the baud rate of the MCP2150's serial port. Matching up the baud rate of the microcontroller to that set by the DIP switches is done in the software that is loaded into PDA Robot's microcontroller. Chapter 7: Programming the PIC16F876 Microcontroller explains this in detail.

Modulation

The data that the MCP2150 UART received (on the TX pin) that needs to be transmitted (on the TXIR pin) will need to be modulated. This modulated signal drives the IR transceiver module. **Figure 5.22** shows the encoding of the modulated signal. Each bit time is comprised of 16-bit clocks. If the value to be transmitted (as determined by the TX pin) is a logic low, then the TXIR pin will output a low level for 7-bit clock cycles, a logic high level for 3-bit clock cycles, or a minimum of 1.6 µs. (see parameter IR121). The remaining 6-bit clock cycles will be low. If the value to transmit is a logic high, then the TXIR pin will output a low level for the entire 16-bit clock cycles. Note: The signal on the TXIR pin does not actually line up in time with the bit value that was transmitted on the TX pin, as shown in **Figure 5.22**. The TX bit value is shown to represent the value to be transmitted on the TXIR pin.

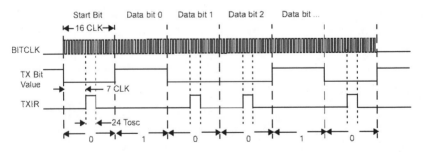

Figure 5.22

MCP2150 data encoding (modulated).

Demodulation

The modulated signal (data) from the IR transceiver module (on RXIR pin) needs to be demodulated to form the received data (on RX pin). Once demodulation of the data byte occurs, the received data are transmitted by the MCP2150 UART (on the RX pin). **Figure 5.23** illustrates the data decoding. Note: The signal on the RX pin does not actually line up in time with the bit value that was received on the RXIR pin, as shown in **Figure 5.23**. The RXIR bit value is shown to represent the value to be transmitted on the RX pin.

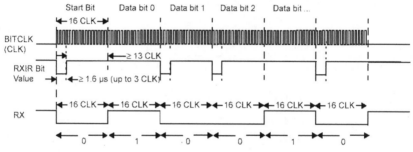

Figure 5.23

MCP2150 data encoding (demodulated).

Minimizing Power

The device can be placed in a low-power mode by disabling the device (holding the EN pin at the low state). The internal state machine is monitoring this pin for a low level. Once this is detected, the device is disabled and enters into a low-power state.

Returning to Device Operation

When disabled, the device is in a low-power state. When the EN pin is brought to a high level, the device will return to the operating mode. The device requires a delay of 1024 TOSC before data may be transmitted or received.

Network Layering Reference Model

Figure 5.24 shows the Open Systems Interconnect (OSI) Network Layering Reference Model. The shaded areas are implemented by the

65

Figure 5.24

OSI layers.

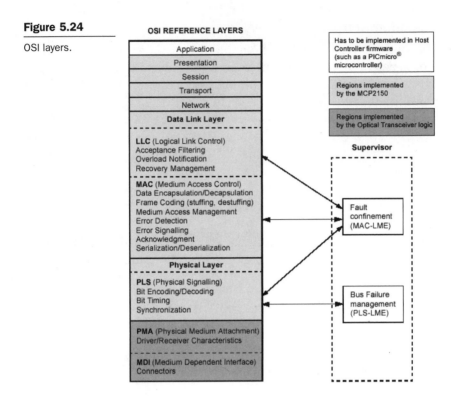

MCP2150; the cross-hatched area is implemented by an IR transceiver. The unshaded areas should be implemented by the host controller.

IrDA Data Protocols Supported by MCP2150

The MCP2150 supports the following required IrDA standard protocols:

- Physical Signaling Layer (PHY)

- Link Access Protocol (IrLAP)

- Link Management Protocol/Information Access Service (IrLMP/ IAS)

The MCP2150 also supports some of the optional protocols for IrDA data. The optional protocols that the MCP2150 implements are:

- Tiny TP

- IrCOMM

The software running on the PDA utilizes all the optional and required protocols supported by the MCP2150 (see Chapters 8 and 9).

Figure 5.25 shows the IrDA data protocol stack and which components are implemented by the MCP2150. The optional IR transceiver for the asynchronous serial IR is the Vishay transceiver described earlier.

Physical Signal Layer (PHY). The MCP2150 provides the following Physical Signal Layer specification support:

- Bidirectional communication.

- Data Packets are protected by a CRC—16-bit CRC for speeds up to 115.2 kbaud.

- Data Communication Rate—9600 baud minimum data rate.

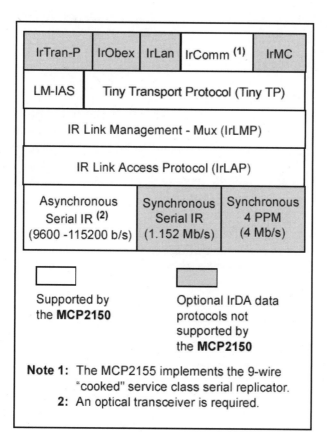

Figure 5.25

MCP2150 IrDA protocol stack.

The following Physical Layer Specification is dependent on the optical transceiver logic used in the application. The specification states:

- Communication range, which sets the end user expectation for discovery, recognition, and performance.

- Continuous operation from contact to at least 1 m (typically 2 m can be reached).

- A low power specification reduces the objective for operation from contact to at least 20 cm (low power and low power) or 30 cm (low power and standard power).

IrLAP. The MCP2150 supports the IrLAP. The IrLAP provides:

- Management of communication processes on the link between devices.

- A device-to-device connection for the reliable, ordered transfer of data.

- Device discover procedures.

- Hidden node handling.

Figure 5.26 identifies the key parts and hierarchy of the IrDA protocols. The bottom layer is the Physical layer, IrPHY. This is the part that converts the serial data to and from pulses of IR light. *IR transceivers can't transmit and receive at the same time.* The receiver has to wait for the transmitter to finish sending. This is sometimes referred to as a "half-duplex" connection. The IR IrLAP provides the structure for

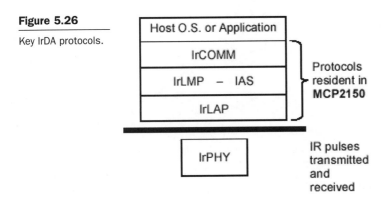

Figure 5.26

Key IrDA protocols.

packets (or "frames") of data to emulate data that would normally be free to stream back and forth.

IRDA Standard Protocol Layers

The IrLAP frame is proceeded by some number of Beginning of Frame characters (BOFs). The value of the BOF is generally 0xC0, but 0xFF may be used if the last BOF character is a 0xC0. The purpose of multiple BOFs is to give the other station some warning that a frame is coming. The IrLAP frame begins with an address byte ("A" field), then a control byte ("C" field). The control byte is used to differentiate between different types of frames and is also used to count frames. Frames can carry status, data, or commands. The IrLAP has a command syntax of its own. These commands are part of the control byte. Last, IrLAP frames carry data. These data are the information (or "I") field. The integrity of the frame is ensured with a 16-bit CRC, referred to as the frame check sequence (FCS). The 16-bit CRC value is transmitted least signification bit (LSB) first. The end of the frame is marked with an end of frame (EOF) character, which is always a 0xC1. The frame structure described here is used for all versions of IrDA protocols used for serial wire replacement for speeds up to 115.2 kbaud.

In addition to defining the frame structure, IrLAP provides the "housekeeping" functions of opening, closing, and maintaining connections. The critical parameters that determine the performance of the link are part of this function. These parameters control how many BOFs are used, identify the speed of the link, how fast either party may change from receiving to transmitting, etc. IrLAP has the responsibility of negotiating these parameters to the highest common set so that both sides can communicate as quickly, and as reliably, as possible. This is done during the handshaking phase when the PDA is connecting to PDA Robot.

IrLMP. The MCP2150 implements the IrLMP. The IrLMP provides:

- Multiplexing of the IrLAP layer. This allows multiple channels above an IrLAP connection.

- Protocol and service discovery via the IAS.

When two devices that contain the IrDA standard feature are connected, generally one device has something to do and the other device has the resource to do it. For example, a laptop may have a job to print and an IrDA standard compatible printer has the resources to print it. In IrDA standard terminology, the laptop is a primary device and the printer is the secondary device. When these two devices connect, the primary device must determine the capabilities of the secondary device to determine if the secondary device is capable of doing the job. This determination is made by the primary device asking the secondary device a series of questions. Depending on the answers to these questions, the primary device may or may not elect to connect to the secondary device.

The queries from the primary device are carried to the secondary device using IrLMP. The responses to these queries can be found in the IAS of the secondary device. The IAS is a list of the resources of the secondary device. The primary device compares the IAS responses with its requirements, and then makes the decision if a connection should be made. For instance, the software running on the PDA queries PDA Robot to see what it identifies itself as, and to see if it will accept the "cooked-wire" service. If it identifies itself as what we are looking for and supports the service, then a connection is made.

The MCP2150 identifies itself to the primary device as a modem. The MCP2150 is not a modem, and the nondata circuits are not handled in a modem fashion.

Link Management-Information Access Service (LM-IAS). The MCP2150 implements the LM-IAS. Each LM-IAS entity maintains an information database to provide:

- Information on services for other devices that contain the IrDA standard feature (Discovery).

- Information on services for the device itself.

- Remote accessing of another device's information base.

This is required so clients on a remote device can find configuration information needed to access a service.

Tiny TP. Tiny TP provides the flow control on IrLMP connections. An optional service of Segmentation and Reassembly can be handled.

IrCOMM. IrCOMM provides the method to support serial and parallel port emulation. This is useful for legacy COM applications, such as printers and modem devices. The IrCOMM standard is just a syntax that allows the primary device to consider the secondary device as a serial device. IrCOMM allows for emulation of serial or parallel (printer) connections of various capabilities.

The MCP2150 (PDA Robot) supports the 9-wire "cooked" service class of IrCOMM. Other service classes supported by IrCOMM are shown in **Figure 5.27**. Note: The MCP2150 identifies itself as a modem to ensure that it is identified as a serial device with a limited amount of memory.

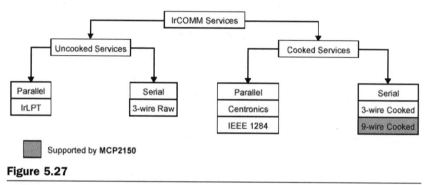

Figure 5.27

Services supported by IrCOMM.

PDA and PDA Robot Handshake: How Devices Connect

When two devices implementing the IrDA standard feature (PDA and PDA Robot) establish a connection using the IrCOMM protocol, the process is analogous to connecting two devices with serial ports using a cable. This is referred to as a point-to-point connection. This connection is limited to half-duplex operation because the IR transceiver cannot transmit and receive at the same time. The purpose of the IrDA protocol is to allow this half-duplex link to emulate, as much as possible, a full-duplex connection. In general, this is done by dividing the data into "packets," or groups of data. These packets can then be sent back and forth, when needed, without risk of collision. The rules of how and when these packets are sent constitute the IrDA protocols. The MCP2150 supports elements of this IrDA protocol to communicate with other IrDA standard compatible devices. When a wired connection is used, the assumption is made that both sides have the same communications parameters and features. A wired connection has no need to identify the other connector because it is assumed that the connectors are properly connected. In the IrDA standard, a connection process has been defined to identify other IrDA compatible devices and establish a communication link. These two devices (PDA and PDA Robot) go through three steps to make this connection. They are:

- Normal disconnect mode (NDM)

- Discovery mode

- Normal connect mode (NCM)

Figure 5.28 shows the connection sequence.

Normal Disconnect Mode (NDM)

When two IrDA standard compatible devices come into range, they must first recognize each other. The basis of this process is that one device has some task to accomplish, and the other device has a resource needed to accomplish this task. One device is referred to as a primary device and the other is referred to as a secondary device. This distinction between primary device and secondary device is important. In our case, the PDA is the primary device and PDA Robot is the secondary. It is the responsibility of the primary device to provide the mechanism to recognize other devices.

So the primary device must first poll for nearby IrDA standard compatible devices. During this polling, the default baud rate of 9600 baud is used by both devices. For example, to print from an IrDA-equipped laptop to an IrDA printer, utilizing the IrDA standard feature, first bring your laptop in range of the printer. In this case, the laptop has something to do and the printer has the resource to do it. The laptop is called the primary device and the printer is the secondary device. Some data-capable cell phones have IrDA standard IR ports. If you used such a cell phone with a PDA, the PDA that supports the IrDA standard feature would be the primary device, and the cell phone would be the secondary device.

When a primary device polls for another device, a nearby secondary device may respond. When a secondary device responds, the two devices are defined to be in the NDM state. NDM is established by the

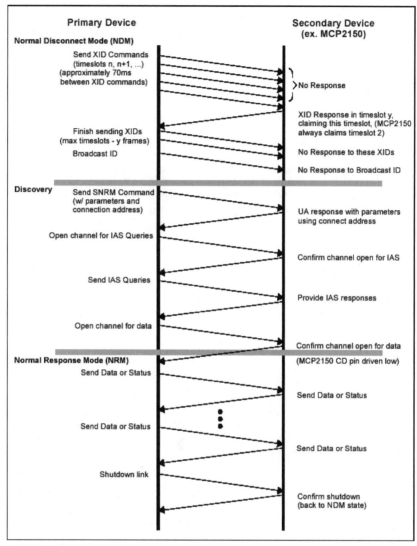

Figure 5.28

Connection sequence.

primary device broadcasting a packet and waiting for a response. These broadcast packets are numbered. Usually 6 or 8 packets are sent. The first packet is number 0, and the last packet is usually number 5 or 7. Once all the packets are sent, the primary device sends an ID packet, which is not numbered. The secondary device waits for these packets, and then responds to one of the packets. The packet it responds to determines the "time slot" to be used by the secondary device. For example, if the secondary device responds after packet number 2, then the secondary device will use time slot 2. If the secondary device responds after packet number 0, then the secondary device will use time slot 0. This mechanism allows the primary device to recognize as many nearby devices as there are time slots. The primary device will continue to generate time slots, and the secondary device should continue to respond, even if there is nothing to do.

During NDM, the MCP2150 handles all of the responses to the primary device (**Figure 5.28**) without any communication with the host controller. The host controller is inhibited by the clear to send (CTS) signal of the MCP2150 from sending data to the MCP2150. Note the following:

- The MCP2150 can only be used to implement a secondary device.

- The MCP2150 supports a system with only one secondary device having exclusive use of the IrDA standard IR link (known as point-to-point communication).

- The MCP2150 always responds to packet number 2. This means that the MCP2150 will always use time slot 2.

- If another secondary device is nearby, the primary device may fail to recognize the MCP2150, or the primary device may not recognize either of the devices. This is not the case with the software developed for the PDAs. I get a list of all secondary devices that respond to the discovery request and look for the identifier for PDA Robot, which is "Generic IrDA" (the default setting of the MCP2150). My printer and PDA Robot always respond, and the software *only* connects to PDA Robot.

Discovery Mode

Discovery mode allows the primary device to determine the capabilities of the MCP2150 (secondary device). Discovery mode is entered once the MCP2150 (secondary device) has sent an XID response to the primary device and the primary device has completed sending the XIDs, and then sends a Broadcast ID. If this sequence is not completed, then a primary and secondary device can stay in NDM indefinitely. When the primary device has something to do, it initiates Discovery. Discovery has two parts. They are:

• Link initialization

• Resource determination

The first step is for the primary and secondary devices to determine, and then adjust to, each other's hardware capabilities. These capabilities are parameters like:

• Data rate

• Turnaround time

• Number of packets without a response

• How long to wait before disconnecting

Both the primary and secondary device begin communications at 9600 baud, which is the default baud rate. The primary device sends its parameters, then the secondary device responds with its parameters. For example, if the primary supports all data rates up to 115.2 kbaud and the secondary device only supports 19.2 kbaud, the link will be established at 19.2 kbaud.

Once the hardware parameters are established, the primary device must determine if the secondary device has the resources it requires. If the primary device has a job to print, then it must know if it is talking to a printer, not a modem or other device. This determination is made using the IAS. The job of the secondary device is to respond to IAS queries made by the primary device. The primary device must ask a series of questions like:

• What is the name of your service?

• What is the address of this service?

• What are the capabilities of this device?

When all the primary device's questions are answered, the primary device can access the service provided by the secondary device. During Discovery mode, the MCP2150 handles all responses to the primary device (see **Figure 5.28**) without any communication with the host controller. The host controller is inhibited by the CTS signal of the MCP2150 from sending data to the MCP2150.

Normal Connect Mode (NCM)

Once discovery has been completed, the primary device and MCP2150 (secondary device) can freely exchange data. The MCP2150 can receive IR data or serial data, but not both simultaneously. The MCP2150 uses a hardware handshake to stop the local serial port from sending data while the MCP2150 is receiving IR data. Both the primary device and the MCP2150 (secondary device) check to make sure that data packets are received by the other without errors. Even when data is required to be sent, the primary and secondary devices will still exchange packets to ensure that the connection has not, unexpectedly, been dropped.

When the primary device has finished, it then transmits the close link command to the MCP2150 (secondary device). The MCP2150 will confirm the close link command, and both the primary device and the MCP2150 (secondary device) will revert to the NDM state. It is the responsibility of the host controller program to understand the meaning of the data received and how the program should respond to it. It is just as if the data were being received by the host controller from a UART. Note: The MCP2150 is limited to a data rate of 115.2 kbaud. Data loss will result if this hardware handshake is not observed. If the NCM mode is unexpectedly terminated for any reason (including the primary device not issuing a close link command), the MCP2150 will revert to the NDM state 10 seconds after the last frame has been received. **Figure 5.28** shows the connection sequence.

MCP2150 Operation

The MCP2150 emulates a null modem connection. The application on the DTE device sees a virtual serial port. This serial port emulation is provided by the IrDA standard protocols. The link between the DTE device and the embedded application is made using the MCP2150. The connection between the MCP2150 and the embedded application is wired as if there were a null modem connection.

The carrier detect (CD) signal of the MCP2150 is used to indicate that a valid IrDA standard IR link has been established between the MCP2150 and the primary device. The CD signal should be monitored closely to make sure that any communication tasks can be completed. The MCP2150 data signaling rate (DSR) signal indicates that the device has powered up, is successfully initialized, and is ready for service. This signal is intended to be connected to the DSR input of the host controller. If the host controller was directly connected to an IrDA standard primary device using a serial cable (the MCP2150 is not present), the host controller would be connected to the primary device's data transfer rate (DTR) output signal. The MCP2150 generates the CTS signal locally because of buffer limitations. Only the transceiver's TXD and RXD signals are carried back and forth to the primary device. The MCP2150 emulates a three-wire serial connection (TXD, RXD, and GND).

The code for the PIC16F876 used in PDA Robot creates a three-wire serial connection with the MCP2150 using the following line of code. See Chapter 7: Programming the PIC16F876 Microcontroller.

```
#use rs232(baud=115200, xmit=PIN_B1, rcv=PIN_B0, stream=PDA)
```

Optical Transceiver

The MCP2150 requires an IR transceiver. The transceiver that we are using is the TFDS4500, as described earlier in this chapter. The transceiver can be an integrated solution. A typical optical transceiver circuit, using a Vishay TFDS4500, is shown in **Figure 5.29**.

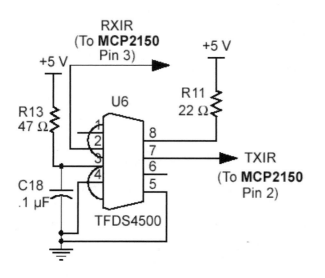

Figure 5.29

Typical transceiver interface to the MCP2150.

Typical Optical Transceiver Circuit

The optical transceiver logic can be implemented with discrete components for cost savings. Care must be taken in the design and layout of the photo-detect circuit, due to the small signals that are being detected and their sensitivity to noise.

MCP2150 Absolute Maximum Ratings

Ambient Temperature under bias . . . . −40°C to +125°C

Storage Temperature. . . . . −65°C to +150°C

Voltage on VDD with respect to VSS . . . . −0.3 V to +6.5 V

Voltage on RESET with respect to VSS . . . . −0.3 V to +14 V

Voltage on all other pins with respect to VSS . . . . −0.3 V to (VDD + 0.3 V)

Total Power Dissipation (1). . . . . 800 mW

Max. Current out of VSS pin . . . . 300 mA

Max. Current into VDD pin . . . . 250 mA

Input Clamp Current, IIK (VI < 0 or VI > VDD) . . . . ±20 mA

Output Clamp Current, IOK (VO < 0 or VO > VDD) . . . . ±20 mA

Max. Output Current sunk by any Output pin . . . . 25 mA

Max. Output Current sourced by any Output pin. . . . . 25 mA

Note 1: Power Dissipation is calculated as follows:

PDIS = VDD x {IDD - Σ IOH} + Σ {(VDD-VOH) x IOH} + Σ(VOL x IOL)

NOTICE: Stresses above those listed under "Maximum Ratings" may cause permanent damage to the device. This is a stress rating only, and functional operation of the device at those or any other conditions above those indicated in the operational listings of this specification is not implied. Exposure to maximum rating conditions for extended periods may affect device reliability.

Figure 5.30 shows the physical layout of the MCP2150 chip used in PDA Robot.

PIC16F876: PDA Robot's Microcontroller

The PIC16F876 is used to send and receive commands from the robot to the PDA, get analog readings from the range finder, and switch the robot's motors on and off. I chose this chip because it is low cost, very fast, can be electronically erased, flashed programmed, and is readily available.

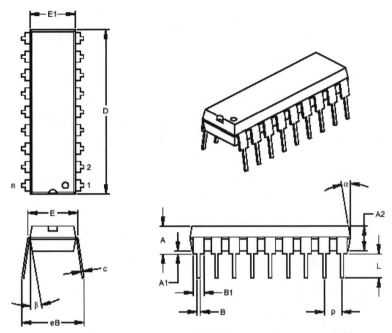

Units		INCHES*			MILLIMETERS		
Dimension Limits		MIN	NOM	MAX	MIN	NOM	MAX
Number of Pins	n		18			18	
Pitch	p		.100			2.54	
Top to Seating Plane	A	.140	.155	.170	3.56	3.94	4.32
Molded Package Thickness	A2	.115	.130	.145	2.92	3.30	3.68
Base to Seating Plane	A1	.015			.038		
Shoulder to Shoulder Width	E	.300	.313	.325	7.62	7.94	8.26
Molded Package Width	E1	.240	.250	.260	6.10	6.35	6.60
Overall Length	D	.890	.898	.905	22.61	22.80	22.99
Tip 10 Seating Plane	L	.125	.130	.135	3.18	3.30	3.43
Lead Thickness	c	.008	.012	.015	0.20	0.29	0.38
Upper Lead Width	B1	.045	.058	.070	1.14	1.46	1.78
Lower Lead Width	B	.014	.018	.022	0.36	0.46	0.56
Overall Row Spacing §	eB	.310	.370	.430	7.87	9.40	10.92
Mold Draft Angle top	α	5	10	15	5	10	15
Mold Draft Angle Bottom	ß	5	10	15	5	10	15

* Controlling Parameter
§ Significant Characteristic
Notes:
Dimensions D and E1 do not include mold flash or protrusions. Mold flash or protrusions shall not exceed
.010" (0.254mm) per side.
JEDEC Equivalent: MS-001
Drawing No. C04-007

Figure 5.30

MCP2150 DIP physical dimensions used in PDA Robot.

The following information about the specifics of this chip from the data sheets explains the details of its inner workings. I highly recommend going to www.microchip.com to download any updates. There is enough information provided in the sheets to write a C or C++ compiler for the chip if you are so inclined. When the sheet explains how the chip does the analog to digital conversions, you could use that information to create one of your own externally with a capacitor. This would allow you to buy a chip that has only digital input/output pins and create the A/D converter yourself. The following summarizes what you need to know. Features include:

- High-performance RISC CPU

- Only 35 single-word instructions to learn

- All single-cycle instructions except for program branches, which are two cycle

- Operating speed: DC—20 MHz clock input DC—200 ns instruction cycle

- Up to 8K × 14 words of FLASH program memory, up to 368 × 8 bytes of data memory (RAM), up to 256 x 8 bytes of EEPROM data memory

- Pinout compatible to the PIC16C73B/74B/76/77

- Interrupt capability (up to 14 sources)

- Eight-level-deep hardware stack

- Direct, indirect, and relative addressing modes

- Power-on Reset (POR)

- Power-up Timer (PWRT) and Oscillator Start-up Timer (OST)

- Watchdog Timer (WDT) with its own on-chip RC oscillator for reliable operation

- Programmable code protection

- Power saving SLEEP mode

- Selectable oscillator options

- Low-power, high-speed CMOS FLASH/EEPROM technology

- Fully static design

- In-circuit serial programming (ICSP) via two pins

- Single 5V in-circuit serial programming capability

- In-circuit debugging via two pins

- Processor read/write access to program memory

- Wide operating voltage range: 2.0 V to 5.5 V

- High sink/source current: 25 mA

- Commercial, industrial, and extended temperature ranges

- Low power consumption:

    – < 0.6 mA typical @ 3V, 4 MHz

    – 20 μA typical @ 3V, 32 kHz

    – < 1 μA typical standby current peripheral features:

- Timer0: 8-bit timer/counter with 8-bit prescaler

- Timer1: 16-bit timer/counter with prescaler, can be incremented during SLEEP via external crystal/clock

- Timer2: 8-bit timer/counter with 8-bit period register, prescaler, and postscaler

- Two capture, compare, PWM modules

    – Capture is 16-bit; max. resolution is 12.5 ns

    – Compare is 16-bit; max. resolution is 200 ns

    – PWM max. resolution is 10-bit

- 10-bit multi-channel analog-to-digital converter

- Synchronous serial port (SSP) with SPI (master mode) and I to the power of 2 C (master/slave)

- Universal synchronous asynchronous receiver transmitter (USART/SCI) with 9-bit address detection

- Parallel slave port (PSP) 8 bits wide, with external RD, WR, and CS controls (40/44-pin only)

- Brown-out detection circuitry for brown-out reset (BOR)

Figure 5.31 shows the pin layout of the chip.

Figure 5.31

PDIP, SOIC

PIC16F876 pin layout.

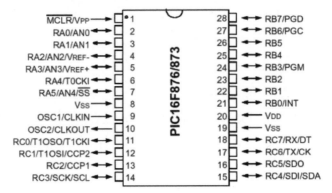

The block diagram in **Figure 5.32** gives you an idea of the chip's inner architecture.

Table 5.4

PIC16F876 Pin Descriptions

Pin Name	Pin #	Pin Type	Buffer Type	Description
OSC1/CLKIN	9	I	ST/ CMOS	Oscillator crystal input/external clock source input.
OSC2/CLKOUT	10	O	—	Oscillator crystal output. Connects to crystal or resonator in crystal oscillator mode. In RC mode, OSC2 pin outputs CLKOUT which has 1/4 the frequency of OSC1, and denotes the instruction cycle rate.
MCLR/VPP	1	I/P	ST	Master Clear (Reset) input or programming voltage input. This pin is an active low RESET to the device.
				PORTA is a bidirectional I/O port.
A0/AN0	2	I/O	TTL	RA0 can also be analog input0.
RA1/AN1	3	I/O	TTL	RA1can also be analog input0.
RA2/AN2/ VREF-	4	I/O	TTL	RA2 can also be analog input2 or negative analog reference voltage.
RA3/AN3/ VREF+	5	I/O	TTL	RA3 can also be analog input3 or positive analog reference voltage.
RA4/T0CKI	6	I/O	TTL	RA4 can also be the clock input to the imer0 timer/counter. Output is open drain type.

(continued on page 84)

Device	Program FLASH	Data Memory	Data EEPROM
PIC16F873	4K	192 Bytes	128 Bytes
PIC16F876	8K	368 Bytes	256 Bytes

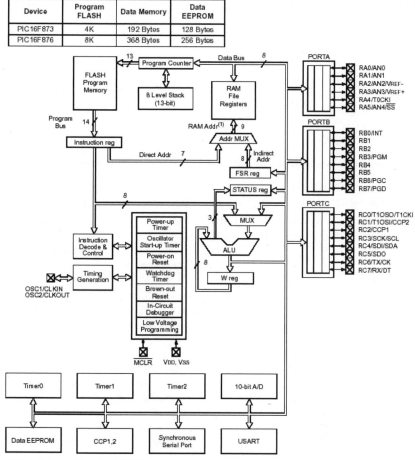

Figure 5.32

PIC16F873 and PIC16F876 block diagram.

83

Table 5.4

PIC16F876 Pin Descriptions (continued)

Pin Name	Pin #	Pin Type	Buffer Type	Description
RA5/SS/AN4	7	I/O	TTL	RA5 can also be analog input4 or the slave select for the synchronous serial port.
				PORTB is a bidirectional I/O port. PORTB can be software programmed for internal weak pull-up on all inputs.
RB0/INT	21	I/O	TTL/ST	RB0 can also be the external interrupt pin.
RB1	22	I/O	TTL	
RB2	23	I/O	TTL	
RB3/PGM	24	I/O	TTL	RB3 can also be the low-voltage programming input.
RB4	25	I/O	TTL	Interrupt-on-change pin.
RB5	26	I/O	TTL	Interrupt-on-change pin.
RB6/PGC	27	I/O	TTL/ST	Interrupt-on-change pin or in-circuit debugger pin. Serial programming clock.
RB7/PGD	28	I/O	TTL/ST	Interrupt-on-change pin or in-circuit debugger pin. Serial programming data.
				PORTC is a bidirectional I/O port.
RC0/T1OSO/T1CKI	11	I/O	ST	RC0 can also be the Timer1 oscillator output or Timer1 clock input.
RC1/T1OSI/CCP2	12	I/O	ST	RC1 can also be the Timer1 oscillator input or Capture2 input/Compare2 output/PWM2 output.
RC2/CCP1	13	I/O	ST	RC2 can also be the Capture1 input/Compare1 output/PWM1 output.
RC3/SCK/SCL	14	I/O	ST	RC3 can also be the synchronous serial clock input/output for both SPI and I2C modes.
RC4/SDI/SDA	15	I/O	ST	RC4 can also be the SPI data in (SPI mode) or data I/O (I2C mode).
RC5/SDO	16	I/O	ST	RC5 can also be the SPI data out (SPI mode).
RC6/TX/CK	17	I/O	ST	RC6 can also be the USART asynchronous transmit or synchronous clock.
RC7/RX/DT	18	I/O	ST	RC7 can also be the USART asynchronous receive or synchronous data.
VSS	8,19	—	P	Ground reference for logic and I/O pins.
VDD	20	—	P	Positive supply for logic and I/O pins.

PORTA and the TRISA Register

PORTA is a 6-bit-wide, bidirectional port. The corresponding data direction register is TRISA. Setting a TRISA bit (= 1) will make the corresponding PORTA pin an input (i.e., put the corresponding output

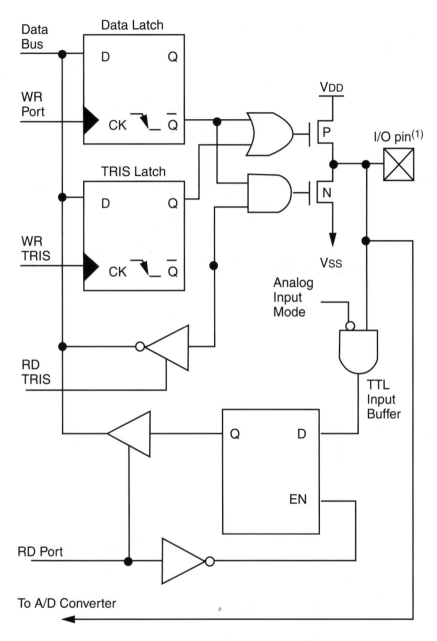

Note 1: I/O pins have protection diodes to VDD and VSS.

Figure 5.33

Block diagram of RA3:RA0 and RA5 pins.

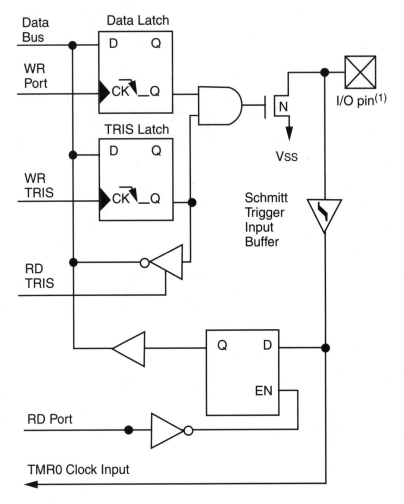

Note 1: I/O pins have protection diodes to Vss only.

Figure 5.34

Block diagram of RA4/TOCK1 pin.

driver in a high-impedance mode). Clearing a TRISA bit (= 0) will make the corresponding PORTA pin an output (i.e., put the contents of the output latch on the selected pin).

Reading the PORTA register reads the status of the pins, whereas writing to it will write to the port latch. All write operations are read-mod-ify-write operations. Therefore, a write to a port implies that the port pins are read, the value is modified, and then written to the port data latch.

Pin RA4 is multiplexed with the Timer0 module clock input to become the RA4/T0CKI pin. The RA4/T0CKI pin is a Schmitt Trigger input and an open drain output. All other PORTA pins have TTL input levels and full CMOS output drivers. Other PORTA pins are multiplexed with analog inputs and analog VREF input. The operation of each pin is selected by clearing/setting the control bits in the ADCON1 register (A/D Control Register1). Note: I/O pin has protection diodes to VSS only.

The TRISA register controls the direction of the RA pins, even when they are being used as analog inputs. The user must ensure the bits in the TRISA register are maintained set when using them as analog inputs.

Chapter 7: Programming the PIC16F876 Microcontroller explains how to set the TRIS registers using a C code macro.

PORTB and the TRISB Register

PORTB is an 8-bit-wide, bidirectional port. The corresponding data direction register is TRISB. Setting a TRISB bit (= 1) will make the corresponding PORTB pin an input (i.e., put the corresponding output driver in a Hi-Impedance mode). Clearing a TRISB bit (= 0) will make the corresponding PORTB pin an output (i.e., put the contents of the output latch on the selected pin).

Three pins of PORTB are multiplexed with the Low Voltage Programming function: RB3/PGM, RB6/PGC, and RB7/PGD.

Each of the PORTB pins has a weak internal pull-up. A single control bit can turn on all the pull-ups. This is performed by clearing bit RBPU (OPTION_REG<7>). The weak pull-up is automatically turned off when the port pin is configured as an output. The pull-ups are disabled on a Power-on Reset.

Four of the PORTB pins, RB7:RB4, have an interrupt on-change feature. Only pins configured as inputs can cause this interrupt to occur (i.e., any RB7:RB4 pin configured as an output is excluded from the interrupton-change comparison). The input pins (of RB7:RB4) are compared with the old value latched on the last read of PORTB. The

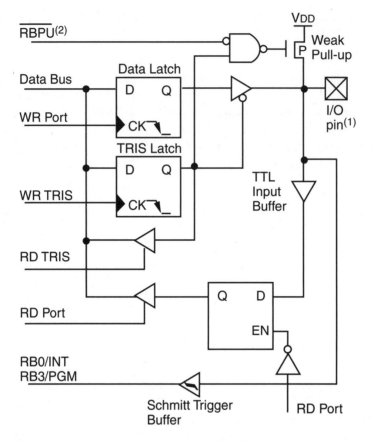

Note 1: I/O pins have diode protection to VDD and Vss.

   2: To enable weak pull-ups, set the appropriate TRIS
   bit(s) and clear the RBPU bit (OPTION_REG<7>).

Figure 5.35

Block diagram of RB3:RB0 pins.

"mismatch" outputs of RB7:RB4 are ORed together to generate the RB Port Change Interrupt with flag bit RBIF (INTCON<0>). This interrupt can wake the device from SLEEP. The user, in the Interrupt Service Routine, can clear the interrupt in the following manner:

- Any read or write of PORTB. This will end the mismatch condition.

- Clear flag bit RBIF.

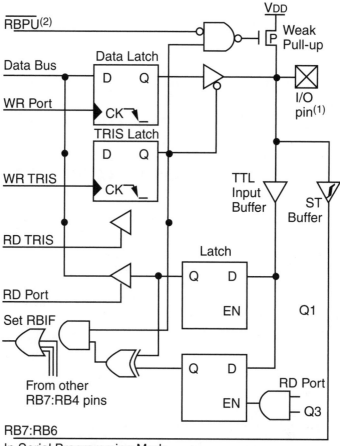

Note 1: I/O pins have diode protection to VDD and VSS.

2: To enable weak pull-ups, set the appropriate TRIS bit(s) and clear the RBPU bit (OPTION_REG<7>).

Figure 5.36

Block diagram of RB7:RB4 pins.

A mismatch condition will continue to set flag bit RBIF. Reading PORTB will end the mismatch condition and allow flag bit RBIF to be cleared. The interrupt-on-change feature is recommended for wake-up on key depression operation and operations where PORTB is only used for the interrupt-on-change feature. Polling of PORTB is not recommended while using the interrupt-on-change feature.

This interrupt-on-mismatch feature, together with software configurable pull-ups on these four pins, allows easy interface to a keypad and make it possible for wake-up on key depression.

PORTC and the TRISC Register

PORTC is an 8-bit-wide, bidirectional port. The corresponding data direction register is TRISC. Setting a TRISC bit (= 1) will make the corresponding PORTC pin an input (i.e., put the corresponding output driver in a Hi-Impedance mode). Clearing a TRISC bit (= 0) will make the corresponding PORTC pin an output (i.e., put the contents of the output latch on the selected pin).

PORTC is multiplexed with several peripheral functions. PORTC pins have Schmitt Trigger input buffers. When the I2C module is enabled, the PORTC<4:3> pins can be configured with normal I2C levels, or with SMBus levels by using the CKE bit (SSPSTAT<6>). When enabling peripheral functions, care should be taken in defining TRIS bits for each PORTC pin. Some peripherals override the TRIS bit to make a pin an output, while other peripherals override the TRIS bit to make a pin an input. Since the TRIS bit override is in effect while the peripheral is enabled, read modify write instructions (BSF, BCF, XORWF) with TRISC as destination, should be avoided. The user should refer to the corresponding peripheral section for the correct TRIS bit settings.

Analog-to-Digital Converter (A/D) Module. The Analog-to-Digital (A/D) Converter module has five inputs for the 28-pin devices and eight for the other devices. The analog input charges a sample and hold capacitor. The output of the sample and hold capacitor is the input into the converter. The converter then generates a digital result of this analog level via successive approximation. The A/D conversion of the analog input signal results in a corresponding 10-bit digital number. The A/D module has high- and low-voltage reference input that is software selectable to some combination of VDD, VSS, RA2, or RA3. The A/D converter has a unique feature of being able to operate while the device is in SLEEP mode. To operate in SLEEP, the A/D clock must be derived from the A/D's internal RC oscillator.

The A/D module has four registers. These registers are:

- A/D Result High Register (ADRESH)

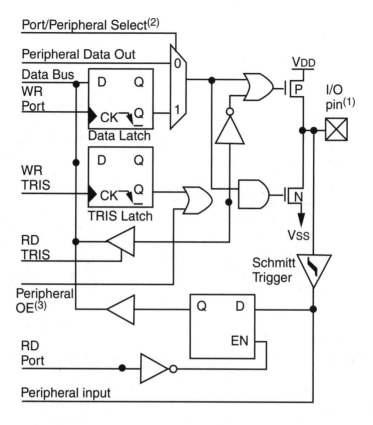

Note 1: I/O pins have diode protection to VDD and VSS.

2: Port/Peripheral select signal selects between port data and peripheral output.

3: Peripheral OE (output enable) is only activated if peripheral select is active.

Figure 5.37

PORTC block diagram (peripheral output override) RC<2.0>, RC<7:5>.

- A/D Result Low Register (ADRESL)

- A/D Control Register0 (ADCON0)

- A/D Control Register1 (ADCON1)

The ADCON0 register controls the operation of the A/D module. The ADCON1 register configures the functions of the port pins. The port

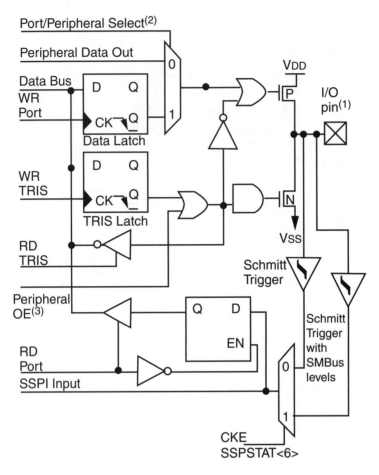

Figure 5.38

PORTC block diagram (peripheral output override) RC<4:3>.

pins can be configured as analog inputs (RA3 can also be the voltage reference), or as digital I/O. Additional information on using the A/D module can be found in the PICmicro Mid-Range MCU Family Reference Manual (DS33023).

Follow these steps when doing an A/D conversion:

1. Configure the A/D module:

   • Configure analog pins/voltage reference and digital I/O (ADCON1).

   • Select A/D input channel (ADCON0).

   • Select A/D conversion clock (ADCON0).

   • Turn on A/D module (ADCON0).

2. Configure A/D interrupt (if desired):

   • Clear ADIF bit.

   • Set ADIE bit.

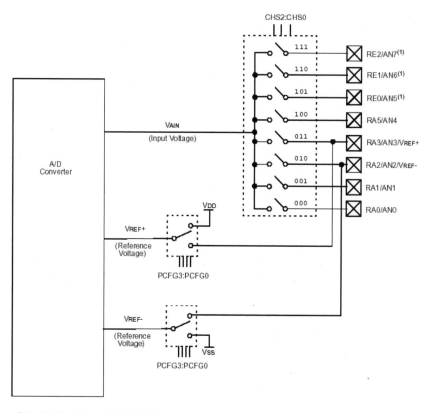

Note 1: Not available on PIC16F873/876 devices.

Figure 5.39

A/D block diagram.

- Set PEIE bit.

- Set GIE bit.

3. Wait for the required acquisition time.

4. Start conversion:

   - Set GO/DONE bit (ADCON0).

5. Wait for A/D conversion to complete, by either:

   - Polling for the GO/DONE bit to be cleared (with interrupts enabled); or

   - Waiting for the A/D interrupt.

6. Read A/D result register pair (ADRESH:ADRESL); clear bit ADIF if required.

7. For the next conversion, go to step 1 or step 2, as required. The A/D conversion time per bit is defined as TAD. A minimum wait of 2TAD is required before the next acquisition starts.

Once again, the C compiler we are using in this project takes care of the preceding steps in a few simple lines of code!

Timer0 Module.   The Timer0 module timer/counter has the following features:

- 8-bit timer/counter

- Readable and writable

- 8-bit software programmable prescaler

- Internal or external clock select

- Interrupt on overflow from FFh to 00h

- Edge select for external clock

Figure 5.40 is a block diagram of the Timer0 module and the prescaler shared with the WDT.

Timer mode is selected by clearing bit T0CS (OPTION_REG<5>). In Timer mode, the Timer0 module will increment every instruction cycle (without prescaler). If the TMR0 register is written, the increment is

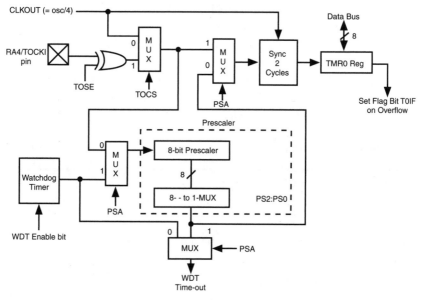

Note: TOCS, TOSE, PSA, PS2:PS0 are (OPTION_REG<5:0>.

Figure 5.40

Block diagram of the Timer0/WDT prescaler.

inhibited for the following two instruction cycles. The user can work around this by writing an adjusted value to the TMR0 register.

Counter mode is selected by setting bit T0CS (OPTION_REG<5>). In Counter mode, Timer0 will increment on either every rising or every falling edge of pin RA4/T0CKI. The incrementing edge is determined by the Timer0 Source Edge Select bit, T0SE (OPTION_REG<4>). Clearing bit T0SE selects the rising edge.

The prescaler is mutually exclusively shared between the Timer0 module and the WDT. The prescaler is not readable or writable.

Timer0 Interrupt.   The TMR0 interrupt is generated when the TMR0 register overflows from FFh to 00h. This overflow sets bit T0IF (INT-CON<2>). The interrupt can be masked by clearing bit T0IE (INT-CON<5>). Bit T0IF must be cleared in software by the Timer0 module Interrupt Service Routine before re-enabling this interrupt. The TMR0 interrupt cannot awaken the processor from SLEEP, since the timer is shut-off during SLEEP.

Using Timer0 with an External Clock. When no prescaler is used, the external clock input is the same as the prescaler output. The synchronization of T0CKI with the internal phase clocks is accomplished by sampling the prescaler output on the Q2 and Q4 cycles of the internal phase clocks. Therefore, it is necessary for T0CKI to be high for at least 2Tosc (and a small RC delay of 20 ns) and low for at least 2Tosc (and a small RC delay of 20 ns).

Prescaler. There is only one prescaler available, which is mutually exclusively shared between the Timer0 module and the WDT. A prescaler assignment for the Timer0 module means that there is no prescaler for the WDT, and vice versa. This prescaler is not readable or writable (see **Figure 5.39**).

The PSA and PS2:PS0 bits (OPTION_REG<3:0>) determine the prescaler assignment and prescale ratio. When assigned to the Timer0 module, all instructions writing to the TMR0 register (e.g., CLRF 1, MOVWF 1, BSF 1, etc.) will clear the prescaler. When assigned to WDT, a CLRWDT instruction will clear the prescaler along with the WDT. The prescaler is not readable or writable.

Note: Writing to TMR0, when the prescaler is assigned to Timer0, will clear the prescaler count, but will not change the prescaler assignment.

The L298 Dual Full-Bridge Driver (PDA Robot Motor Controller)

- Operating supply voltage up to 46 V

- Total DC current up to 4 A

- Low saturation voltage

- Over temperature protection

- Logical "0" Input voltage up to 1.5 V (high noise immunity)

Figure 5.41 shows two of the three available packages that the L298 comes in. In this project, we are using the vertical package shown on the left.

Figure 5.41

L298 packages.

Multiwatt15

PowerSO20

Description

The L298 is an integrated monolithic circuit in 15-lead Multiwatt and PowerSO20 packages. It is a high-voltage, high-current, dual full-bridge driver designed to accept standard TTL logic levels and drive inductive loads such as relays, solenoids, DC, and stepping motors. Two enable inputs are provided to enable or disable the device independently of the input signals. The emitters of the lower transistors of each bridge are connected together, and the corresponding external terminal can be used for the connection of an external sensing resistor.

An additional supply input is provided so that the logic works at a lower voltage, as is the case in PDA Robot. The logic supply comes from the 5 V regulator on the main board via the ribbon connector and the power supply, which drives the motors directly from the 6 V bat-

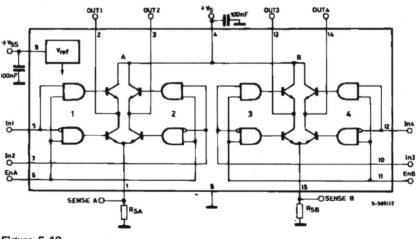

Figure 5.42

L298 block diagram.

tery pack. The grounds are, and must be, connected (common ground). The circuit's block diagram is shown in **Figure 5.42**.

Maximum Ratings. The maximum ratings are shown in **Figure 5.43**. I really like this chip because it will shut down if it is overloaded and becomes hot. It is a nasty sight (and smell) seeing a smoke plume when an overloaded component like a transistor melts down. The L298 can handle a respectable load for its compact size (3 amps, 25 watts). It might be overkill for the motors it drives in this project, but it also means that you can connect much more powerful motors if you decide to change the design. The enable feature and the 2-pin logic lines (with a wide voltage range of –0.3 to 7 V) per side makesa great logic interface. ST microelectronics and Protel provide the footprint and profile for use in the Protel 98 and Protel DXP circuit design programs allowing you to simply drop the chip into your circuit design.

Figure 5.44 a picture of the package we are using in the project showing the pin layout. The short pins are set forward, and the longer are to the back of the chip.

Table 5.5

Pin Descriptions

Pins	Name	Function
1,15	Sense A; Sense B	Between this pin and ground is the sense resistor connected to control the current of the load.
2,3	Out 1; Out 2	Outputs of the Bridge A; the current that flows through the load connected between these two pins is monitored at pin 1.

(continued on next page)

Symbol	Parameter	Value	Unit
V_S	Power Supply	50	V
V_{SS}	Logic Supply Voltage	7	V
V_i, V_{en}	Input and Enable Voltage	–0.3 to 7	V
I_O	Peak Output Current (each Channel) – Non Repetitive (t = 100µs) –Repetitive (80% on –20% off; t_{on} = 10ms) –DC Operation	3 2.5 2	A A A
V_{sens}	Sensing Voltage	–1 to 2.3	V
P_{tot}	Total Power Dissipation (T_{case} = 75°C)	25	W
T_{op}	Junction Operating Temperature	–25 to 130	°C
T_{stg}, T_j	Storage and Junction Temperature	–40 to 150	°C

Figure 5.43

Maximum ratings.

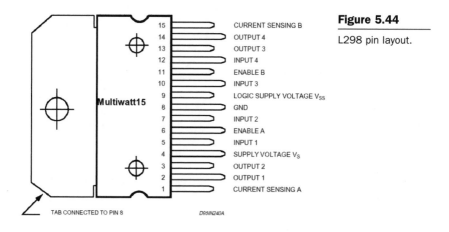

Figure 5.44

L298 pin layout.

Table 5.5

Pin Descriptions (continued)

Pins	Name	Function
4	Vs	Supply Voltage for the Power Output stages. A noninductive 100nF capacitor must be connected between this pin and ground.
5,7	Input 1; Input 2	TTL Compatible Inputs of the Bridge A.
6,11	Enable A; EnableB	TTL Compatible Enable Input: the L state disables the bridge A (enable A) and/or the bridge B (enable B).
8	GND	Ground.
9	VSS	Supply Voltage for the Logic Blocks. A100nF capacitor must be connected between this pin and ground.
10,12	Input 3; Input 4	TTL Compatible Inputs of the Bridge B.
13,14	Out 3; Out 4	Outputs of the Bridge B. The current that flows through the load connected between these two pins is monitored at pin 15.

Figure 5.45 shows how to wire one side of the chip for bidirectional motor control. This is how the chip is wired in PDA Robot. Pins 10 and 12 are connected to Port B pins on the PIC16F876 that have been configured through the C code as outputs (see Chapter 7: Programming the PIC16F876 Microcontroller). In PDA Robot, the sense pins 1 and 15 are tied to the ground. We can feed this into one of the analog pins on the PIC16F876 and determine the current draw on the motors (explained below). If the motor is drawing too much current, shut it down. You can experiment with this. A command could be sent to the

Figure 5.45

Bidirectional motor control. (C = 1 and D = 0) Forward, (C = 0 and D = 1) Reverse, (C = D) Fast Motor Stop.

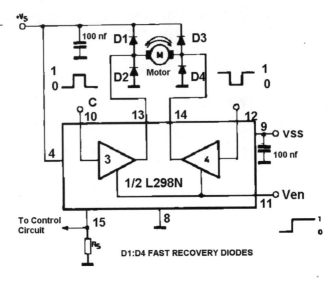

D1:D4 FAST RECOVERY DIODES

Figure 5.46

Paralleled channels for high current.

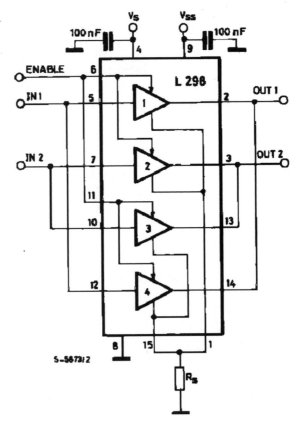

robot to retrieve and forward this information to the PDA (like the range-finder information), where it can be displayed and analyzed. We could determine the speed of PDA Robot based on the current draw after calibrating on a hard, flat surface. This is not a very accurate method of determining the speed and distance traveled, but it will give you a good estimate. Things like the incline and traction will affect the accuracy.

For higher currents, outputs can be paralleled. Take care to parallel channel 1 with channel 4 and channel 2 with channel 3. **Figure 5.46** shows how to accomplish this.

Power Output Stage.   The L298 integrates two power output stages (A; B). The power output stage is a bridge configuration, and its outputs can drive an inductive load in common or differential mode, depending on the state of the inputs. The current that flows through the load comes out from the bridge at the sense output. An external resistor (RSA; RSB) allows one to detect the intensity of this current.

Input Stage.   Each bridge is driven by means of four gates, the input of which are In1; In2; EnA and In3; In4; EnB. The In inputs set the bridge state when the En input is high; a low state of the En input inhibits the bridge. All the inputs are TTL compatible.

Suggestions.   A noninductive capacitor, usually of 100 nF, must be foreseen between both Vs and Vss to ground as near as possible to GND pin. When the large capacitor of the power supply is too far from the IC, a second smaller one must be near the L298. The sense resistor, not of a wire wound type, must be grounded near the negative pole of Vs that must be near the GND pin of the IC. Each input must be connected to the source of the driving signals by means of a very short path.

Turn on and turn off: Before you can turn on the supply voltage and to turn it off; the enable input must be driven to the low state.

Applications.   The external bridge of diodes D1 to D4 is made of four fast recovery elements (trr 3 200 n) that must be chosen from a VF as low as possible at the worst case of the load current. The sense output voltage can be used to control the current amplitude by chopping the inputs, or to provide overcurrent protection by switching the enable input to low.

The brake function (Fast motor stop) requires that the absolute maximum rating of 2 amps must never be overcome. When the repetitive peak current needed from the load is higher than 2 Amps, a paralleled configuration can be chosen.

An external bridge of diodes is required when inductive loads are driven and when the inputs of the IC are chopped; Schottky diodes are preferred. This solution can drive until 3 amps in DC operation and until 3.5 amps of a repetitive peak current. The L298 is great for driving a stepper motor. **Figure 5.47** shows how this is accomplished when the current is controlled by a L6506.

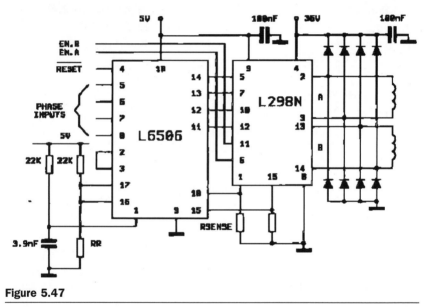

Figure 5.47

Two phase bipolar stepper motor control circuit by using the current controller L6506.

The GP2D12 IR Range Finder

The GP2D12 is a low-cost, short-range IR alternative to ultrasonic range-finding systems. Usable detection range is 10 cm to 80 cm (approx. 4" to 31.5"). The IR Object Detection System consists of the Sharp GP2D12 Distance Measuring Sensor. The GP2D12 is a compact, self-contained IR ranging system incorporating an IR transmitter, receiver, optics, filter, detection, and amplification circuitry. The unit is highly resistant to ambient light and nearly impervious to variations in the surface reflectivity of the detected object.

Unlike many IR systems, it has a fairly narrow field of view, making it easier to get the range of a specific target. The field of view changes with the distance to an object, but is no wider than 5 cm (2.5 cm either side of center) when measuring at the maximum range. One negative about this range finder is its starting range of 10 cm. **Figure 5.48** shows the physical dimensions of the range finder and its connector (www.hvwtech.com).

■ Outline Dimensions (Unit : mm)

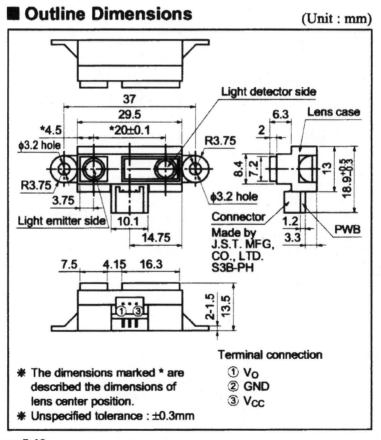

Figure 5.48

Physical dimensions of the range finder.

ABSOLUTE MAXIMUM RATINGS (TA=25 °C, Vcc=5V)			
Parameter	Symbol	Rating	Unit
Supply Voltage	V_{cc}	−0.3 to +7	V
Output Terminal Voltage	V_0	−0.3 to Vcc +0.3	V
Operating Temperature	T_{opr}	−10 to +60	°C
Storage Temperature	T_{stg}	−40 to +70	°C

Figure 5.49

Maximum ratings.

The sensor unit may be mounted using the bracket provided. The black foam should be applied to the bottom of the bracket using the sticky side of the foam, and then the black "snap rivet" is pushed through the large center hole on the bracket. This snap rivet has been chosen to allow the bracket and foam to be mounted on a standard 0.062" PCB. A 13/64" hole is required in the PCB for the snap rivet.

Connecting to the Sensor

A custom cable assembly is included with the kit. The miniature connector is keyed so that it may only be inserted one way: 1 Vcc Red + 5 V DC, 2 GND Black Ground, 3 Vout Blue Input pin of microcontroller

Operation

The GP2D12 makes continuous analog measurements. It does not require a trigger to initiate a measurement. The distance to an object is returned as an analog voltage level. After reading the voltage level produced, a threshold can be set or a distance calculated. By attaching the cabling to a suitable

The analog-to-digital converter or microcontroller with onboard A/D can be incorporated into many systems.

Calibration

The calibration of the module is dependent on how the data are used in your code. For threshold-type applications, calibration involves determining the distance required and measuring the voltage at that distance, allowing for some variations in measurement. In distance measuring applications the relation between voltage level and distance is nonlinear; either a "look-up" table or a suitable calculation

Figure 5.50

Average distance
versus voltage.

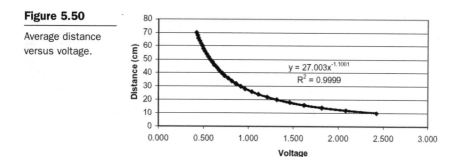

must be determined. The voltage levels representing distance will vary slightly from unit to unit. A small survey of randomly selected devices was conducted and data gathered are shown in **Figure 5.50**. The columns Distance and Average Voltage in the sample data provided can be used as a look-up table.

Using the average of the voltage measurements for the four samples, the following graph was produced. The data points indicate the average values, and the line shows the best fit equation calculated.

The equation derived that best fits the average voltages is given as: Distance (cm) = 27 × (Voltage) −1.1. This equation can be used for calculating the distance to an object by simply entering the voltage measured and calculating the distance in centimeters. The preceding formula is provided for reference only; while it is shown to be quite accurate, part-to-part variation must be considered.

Ambient Light

Tests have shown the GP2D12 to be highly immune to ambient light levels. Incandescent, fluorescent, and natural light do not appear to bother it. The only instance where we were able to get it to falsely measure was when a flashlight was pointed directly into the sensor's receiver; even a few degrees off center is enough for the sensor to ignore it.

IR Light

The GP2D12 uses a modulated IR beam to guard against false triggering from the IR component of incandescent, fluorescent, and natural light. Tests with several kinds of IR remote controls have shown that even with two or three remotes pointed at the GP2D12, the unit still functions normally.

Laser Light

Tests with a laser pointer had results similar to those with the flash-light; only a beam aimed straight into the sensor's receiver would cause a false reading. If the beam comes from even a few degrees off center, it has no effect.

Operation

The GP2D12 uses an array of photo diodes (called a position sensitive detector, or PSD) and some simple optics to detect distance. An IR diode emits a modulated beam; the beam hits an object and a portion of the light is reflected back through the receiver optics and strikes the PSD.

CAUTION: The GP2D12 is a precision device. Do not attempt to open the unit. Doing so will ruin the delicate alignment of the optics. If you want to open one up, by all means do so, but realize beforehand that it may not function properly afterward. A block diagram of the GP2D12 is shown in **Figure 5.51**.

Overall I found this to be an average range finder for PDA Robot. I found that the 10-cm starting range and the narrow beam lead to lim-itations. I will describe them in the chapters on programming the PIC16F876 and PDAs. I would recommend looking into a sonar range finder.

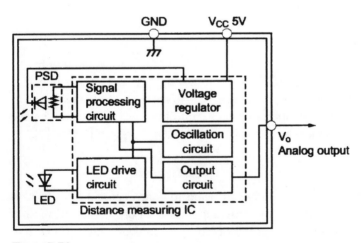

Figure 5.51

GP2D12 block diagram.

6

Building PDA Robot

This chapter explains step-by-step how to create the electrical and mechanical components of PDA Robot.

Creating the Circuit Board

I created the circuit board using the M.G. Chemicals system. The M.G. Chemicals system allows you to make your own circuit boards quickly and easily. It is perfect for prototyping, hobbyists, and educational applications. Technicians will be impressed with the high resolution, while amateurs will be impressed with the simplicity of the system.

I purchased the Photofabrication Kit 416-K to create the PDA Robot circuit board. It includes the following:

- One 3" × 5" cat. #603 presensitized single-sided PCB

- One 4" × 6" cat. #606 presensitized single-sided printed circuit board (PCB)

- One 6" × 6" cat. #609 presensitized single-sided PCB

- One 475 ml bottle cat. #418 developer

- One 475 ml bottle cat. #415 ferric chloride

- Two cat. #416-S foam brushes

- Plastic development tray

- Rubber gloves

- Instruction sheet

Figure 6.1 shows everything that is included in the kit.

Figure 6.1

Contents of the
Photofabrication Kit
416-K.

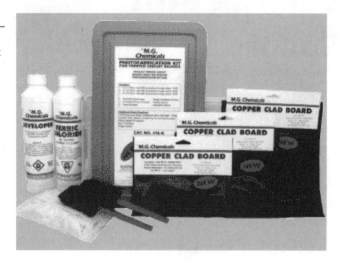

Positive Photofabrication Process Instructions

Setup. Protect surrounding areas from developer and etchant splashes. Plastic is ideal for this. Work under safe light conditions. A 40 W incandescent bulb works well. Important: Do not work under fluorescent light. If you do so, you will expose the board, making it unusable. Just prior to exposure, remove white protective film from the presensitized board. Peel it back carefully.

Exposing Your Board. For best results, use M.G. Chemicals cat. #416-X exposure kit; however, any inexpensive lamp fixture that will hold two or more 18" fluorescent tubes is suitable.

Place the presensitized board with the copper side toward the exposure source. Lay positive film artwork onto the presensitized copper side of the board and position as desired. Place the artwork printed side down to prevent light leakage through the side of the transparency. Artwork should have been produced by a 600 dpi or better printer.

Figure 6.2

Fluorescent
exposure.

Use a glass weight to cover the artwork, ensuring that no light will pass under traces (approx. 3 mm glass thickness or greater works best). Use a 10-minute exposure time at a distance of 5".

The artwork in **Figure 6.3** needs to be reproduced on a transparency and placed on the presensitized "green" surface of the circuit board. **To do this, either scan the artwork and print, make a high-quality photocopy, or download the file from www.pda-robotics.com and print using a photo editor.** From the printer options, set the quality to its highest possible setting. I recommend checking the leads on the components to ensure that the drill holes are the correct size and every hole lines up. **Important:** You must print the image at 100%. If your printer settings are not correct, the components will not fit. Watch out for the components themselves. I found that the higher-priced components fit perfectly, but with some of the less-expensive components, the pad and hole sizes on the artwork may need to be enlarged or the leads filed or crimped. This happened with the voltage regulators and L298 chips. Variations from manufacturer to manufacturer will occur. To increase the hole sizes, simply load the image into an image editor like Paintbrush, and draw in white space after increasing the size of the pad. Be careful when expanding the sizes. You don't want any of the traces to touch each other, and it's good to leave as much space as possible.

After printing the artwork on a good-quality transparency, cut it out using a utility knife or scissors and put it on the presensitized side after carefully peeling the protective cover off (see **Figures 6.3** and **6.4**).

Note: Ensure that the printing on the board in not reversed when placing on the presensitized side. The lettering "PDA Robotics" should be shown as printed normally, not reversed.

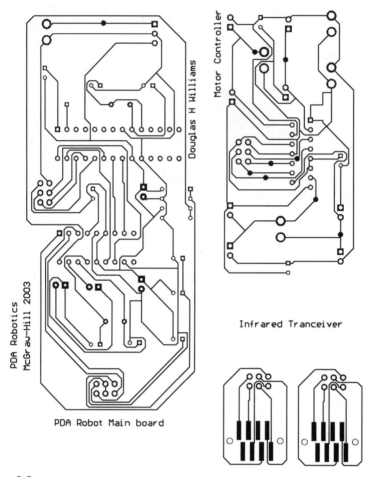

Figure 6.3

Artwork for the circuit board.

Important: Be sure that no fluorescent lights are on anywhere nearby when doing this. Place a clear glass or acrylic weight over the board and transparency and place under the fluorescent light source. Expose the board.

Developing Your Board. The development process removes any photoresist that was exposed through the film positive to ultraviolet light.

Warning: Developer contains sodium hydroxide and is highly corrosive. Wear rubber gloves and eye protection while using it. Avoid con-

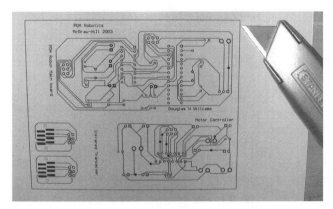

Figure 6.4

Cutting out the circuit board artwork.

tact with eyes and skin. Flush thoroughly with water for 15 minutes if it is splashed in the eyes or on the skin.

Using rubber gloves and eye protection, dilute one part M.G. cat. #418 developer with 10 parts of tepid water (weaker is better than stronger) in a plastic tray. Immerse the board copper side up into the developer, and you will quickly see an image appear while you are lightly brushing the resist with a foam brush. This should be completed within one to two minutes. Immediately neutralize development action by rinsing the board with water. The exposed resist must be removed from the board as soon as possible. When you are done with the developing stage, the only resist remaining will be covering what you want your circuit to be. Completely remove the rest.

Note: Ensure that the mixture of water and developer is mixed thoroughly. If it is not, the traces may wash away when the board comes in contact with a pocket of highly concentrated developer.

Etching Your Board.   For best results, use the 416-E Professional Etching Process Kit or 416-ES Economy Etching Kit. The most popular etchant is ferric chloride, M.G. cat. #415, an aqueous solution that dissolves most metals. Use this solution undiluted, and be sure to completely cover your board.

Warning: This solution is normally heated up during use, generating unpleasant and caustic vapors. It is very important to have adequate ventilation. Use only glass or plastic containers. Keep out of reach of children. It may cause burns or stain. Avoid contact with skin, eyes, or clothing. Store in a plastic container. Wear eye protection and rubber gloves.

Directions: If you use ferric chloride cold, it will take a long time to etch the board. To speed up etching, heat the solution. A simple way of doing this is to immerse the ferric chloride bottle or jug in hot water, adding or changing the water to keep it hot. A thermostat-controlled crock pot is also an effective way to heat ferric chloride, as are thermostatically controlled submersible heaters—glass-enclosed such as an aquarium heater. An ideal etching temperature is 50°C (120°F). Be careful not to overheat it. The absolute maximum working temperature is approximately 57°C (135°F). The warmer the etch solution, the faster the boards will etch. Ferric chloride solution can be used over and over again, until it becomes saturated with copper. As the solution becomes more saturated, the etching time will increase. Agitation assists in removing unwanted copper faster. This can be accomplished by using air bubbles from two aquarium air wands with an aquarium air pump. Do not use aquarium "air stone."

The etching process can be assisted by brushing the unwanted resist with a foam brush while the board is submerged in the ferric chloride. I found that rocking the board back and forth in the pan by holding it by the edges with rubber gloves on works well. The ferric chloride can be kept warm by placing the glass basin on a stove element set to low. Turn on the fume hood to expel any fumes.

After the etching process is completed, wash the board thoroughly under running water. Do not remove the remaining resist protecting your circuit or image. It protects the copper from oxidation.

Figure 6.5

Circuit board ready for etching.

Removal of resist is not necessary when soldering components to the board. By leaving the resist on, you protect the circuit from oxidation. Tin plating the board is not necessary. In soldering, the heat disintegrates the resist underneath the solder, resulting in an excellent bond.

Drilling Out the Circuit Board.   Once the board is dry, drill out the holes using the appropriate drill bits. Be sure that the bit is in straight and that you hit the hole dead center. **Figure 6.6** shows the main board being drilled out. **Figure 6.7** shows the ribbon connector holes on a transponder circuit board.

Cutting the Board.   The board can be cut into the three sections by repeatedly scoring with a utility knife, using a hacksaw, or using a band saw. **Note:** Only one of the infrared (IR) transceiver boards is required for this project. **Figure 6.8** shows where the board should be cut.

Placing and Soldering the Components.   The boards are now ready to have the components soldered into place. The components go onto the topside of the board (opposite side from the traces), with the exception of the TFDS4500 on the transceiver circuit board.

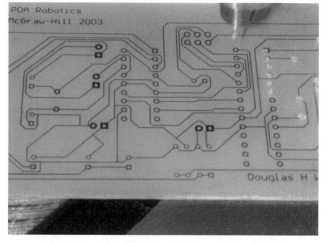

Figure 6.6

Drilling the main board.

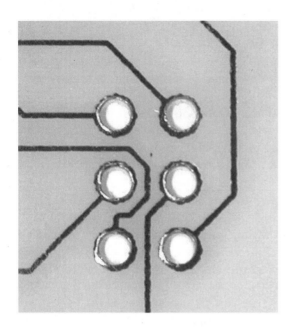

Figure 6.7

Dead center drill.

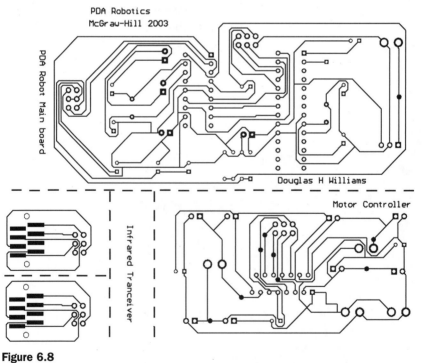

Figure 6.8

Cut the boards on the dotted lines.

Parts Lists

The Main Board.   Parts for the main board include:

- One PIC16F876 microcontroller

- One 28-pin DIP IC socket (or 2 18-pin DIP IC sockets with one cut down)

- One MCP2150 IrDA protocol chip

- One 18-pin DIP IC socket

- One L7805ACV voltage regulator

- One 8-pin DIP switch

- One 11.0592 MHz crystal

- One 20.0000 MHz crystal

- One 3.9 mm (.156") Molex wire connectors

- One Molex 3.9 mm 2P header with ramp connects

- Two 6-post 2.5 mm DIP headers

- Two IDC6F DIP connector with key

- One Red LED

- Three 1 K resistor

- Two 47R 1/4 W resistors

- One 4.7 UF tantalum capacitor

- Six 22 pF capacitors

- One 1-pin

- 8" of six-wire ribbon cable

The Motor Controller.   Parts for the motor controller include:

- One L298N dual bridge driver

- Four 3.9 mm (.156") Molex wire connectors

- Four Molex 3.9 mm 2P header with ramp connects

- Two 6-post 2.5 mm DIP headers

- Two IDC6F DIP connector with key

- Eight 4007 746 diodes

- Three .1 UF capacitors (or higher)

The IR Transceiver.   Parts for the IR transceiver include:

- One TFDS4500

- One 6-post 2.5 mm DIP headers

- One IDC6F DIP connector with key

- 6" of six-wire ribbon cable

Range Finder and Attachments.   Parts for the range finder and attachments include:

- One GP2D12 distance measuring system with cable and attachments (AIRRS @ www.hvwtech.com)

- Two .156" wire connectors

The Body.   Parts for the body include:

- Aluminum: 8" × 6" × 1/16" (main platform)

- Aluminum: 7" × 5 1/4" (top platform) × 1/16"

- Aluminum: 1" × 1/2" × 1/4" (accessory mount)

- Two Tamiya six-speed geared motors (www.hvwtech.com)

- Three Tamiya wheel sets

- Four 1" L-brackets

- Five 2" 4-40 hex spacers

- Eight 1/2" 4-40 hex spacers

- One 9 V battery connector

- One 6 V battery pack (4 × 1.5 V AA)

- 6" of Velcro with self-basting adhesive (secure batteries)

- 1' of double-sided Velcro (secure PDA)

- Package of 50 4-40 1/4" nuts, bolts, and washers.

Placing and Soldering the Main Board Components

Figures **6.9** to **6.11** show the placement of the parts. The following numbers correspond to those on the main circuit board. Place and solder the parts.

1. Molex 3.9 mm 2P header with ramp connects to Molex 156" (3.9 mm) wire connector

2. L7805ACV voltage regulator

3. 22 pF capacitors

4. 1 K resistors

5. Red LED

6. 47R 1/4 W resistors

7. 4.7 UF tantalum capacitor

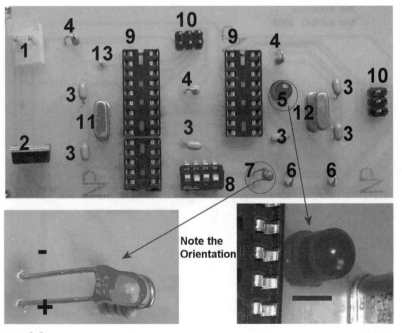

Figure 6.9

Main board parts placement.

8. 8-pin DIP switch

9. 18-pin DIP IC socket (Note the PIC16F876 is using two—one cut down)

10. 6-post 2.5 mm DIP headers

11. 20.0000 MHz crystal

12. 11.0592 MHz crystal

13. 1-pin header for analog input

It is good practice to check the conductance after soldering a component to the board. This ensures that electricity will flow between the points on the circuit and with little resistance. If conductivity is poor, it means that the solder joint is poor and should be redone. To check for conductivity, set the multimeter to RX 1 KΩ, and touch one probe on the solder weld and the other on a trace to which it is connected.

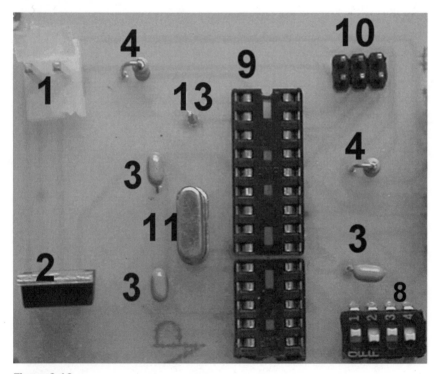

Figure 6.10

Enlarged view of left side of main board.

Figure 6.11

Enlarged view of right side of main board.

The needle on the meter should "spike" to the right, showing zero resistance. **Figure 6.12** shows the meter set to RX 1 KΩ, with the leads crossed and the needle to the far right, indicating that there is no resistance and that the meter is working properly. **Figure 6.13** shows testing the conductivity of the solder connections.

Figure 6.12

Setting the meter for conductivity testing.

Figure 6.13

Testing the solder
connections on a
prototype circuit.

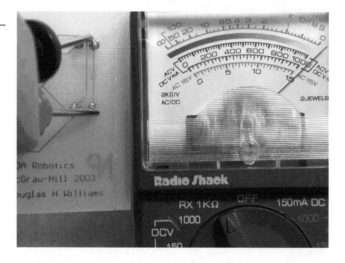

To ensure a good solder joint, keep the tip of the iron clean. Buy high-quality fairly thin solder, and ensure that the iron is hot. Clean the tip after soldering two or three joints.

Placing and Soldering the Motor Controller Components

Figures 6.14 to **6.16** show the placement of the parts on the motor controller circuit. The following numbers correspond to those on the motor controller circuit board. Ensure that the diodes are oriented correctly, as shown in the figure.

1. Molex 3.9 mm 2P headers with ramp connects to Molex .156" wire connectors

2. .1 UF capacitors (or higher)

3. 4007 746 diodes

4. 6-post 2.5 mm DIP headers

5. L298N dual bridge driver

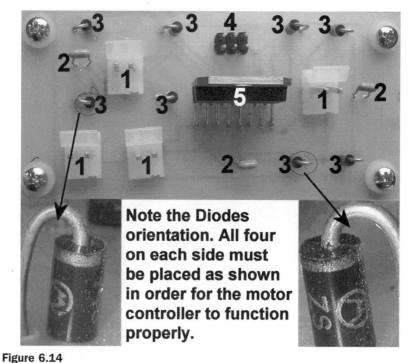

Note the Diodes orientation. All four on each side must be placed as shown in order for the motor controller to function properly.

Figure 6.14

Parts placement on the motor controller circuit board.

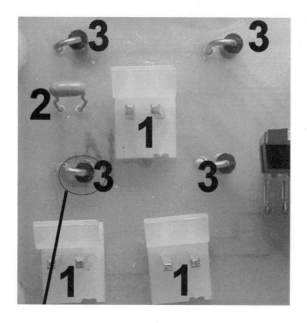

Figure 6.15

Close-up of left side of the motor controller.

Figure 6.16

Close-up of right side of the motor controller.

The Infrared Transceiver

Solder the 6-post 2.5 mm DIP header to the board normally, with the long pins on the top of the board. Position the TFDS4500 on the pads on the bottom of the board, ensuring that the middle of the transceiver is centered over the middle of the pads. Solder or epoxy the pins to the pads being careful to not short any of the pads. Ensure that you are using a good conductive epoxy. **Figure 6.17** shows the TFDS4500 lined up and ready for the epoxy or solder to be applied. **Note:** If using

Figure 6.17

Close-up of the TFDS4500 ready to be soldered or epoxyed to the board.

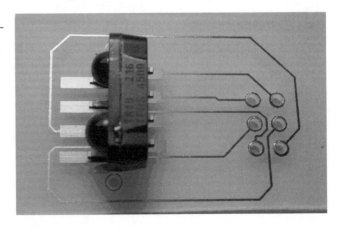

epoxy, gently scrape off the photoresist (which protects the pads from corrosion) in order to achieve a good contact. A small flathead screwdriver works well for this. Once the solder or epoxy has set, it is a good idea to cement the backside of the transceiver with a regular nonconducting epoxy.

Set the boards aside until ready to drill the mounting holes. I recommend putting them in a static-proof bag. We will mount the boards to the craft once the other steps, such as creating the ribbon cables and drilling the holes in the support pieces, etc., are done.

The Power Connectors

The Battery Packs

To prepare the power connectors for the battery packs, motors, and the IR range finder, you will need to solder the Molex .156" (3.9 mm) wire connectors and slide them into the plastic moldings provided. **Figure 6.18** shows the connectors of the battery packs. **Note:** the ground wire is always inserted on the left side of the connector. You may want to solder on/off switches between one of the leads. I simply plug and unplug the power connectors to the posts to turn the craft on or off.

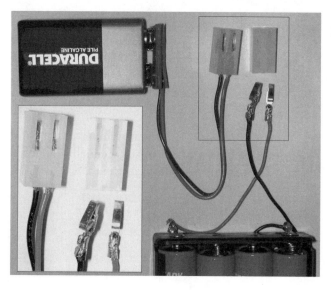

Figure 6.18

The power connections.

The IR Range Finder

The connector that comes with the Sharp GP2D12 needs to have the power leads connected to the 3.9 mm Molex wire connector as well. Solder the black and red wires to the inserts the same way as the battery leads, with the ground wire on the left. The blue wire on the connector goes to the analog input. I improvised a connector for the solitary analog input pin of the PIC16F876 by using a 3.9 mm connector turned around with the end that normally has the wire soldered to it, crimped to fit the pin. This works well because the connector is secured to the pin by the flexible metal tab. **Figure 6.19** shows the soldered connections.

Figure 6.19

The IR range finder connections. A: Positive (red), B: Ground (black), C: Analog line (blue).

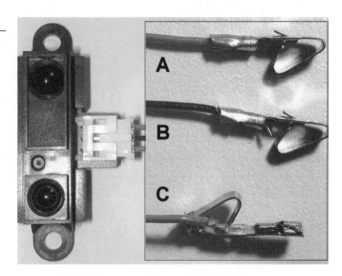

Figure 6.20 shows the improvised connector snug on the PIC16F876 analog input pin.

The two motors will also need to have the Molex power connectors fastened. But first, we must assemble the gear boxes and drill the holes that the wires will feed though from the bottom of the PDA Robot.

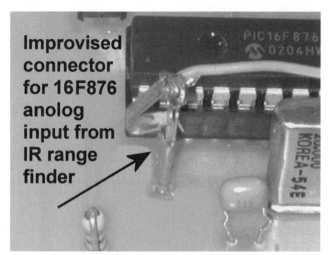

Improvised
connector
for 16F876
anolog
input from
IR range
finder

Figure 6.20

The IR analog input
connector.

Cutting the Aluminum Pieces and Drilling the Holes

Cut the bottom plate (main platform) into an 8" × 6" piece. Cut the top plate that is suspended on four hex spacers to 7" and 5-1/4". Drill out the holes, as outlined in **Figures 6.21** and **6.22**.

- Aluminum: 8" × 6" × 1/16" (main platform)

- Aluminum: 7" × 5-1/4" (top platform) × 1/16"

- Aluminum: 1" × 1/2" × 1/4" (accessory mount)

- Two Tamiya six-speed geared motors (www.hvwtech.com)

- Three Tamiya wheel sets

- Four 1" L-brackets

Mount the hex brackets on the top of the platform. Mount the motors, wheel brackets, and range finder on the bottom. **Figure 6.23** shows the underside with the motors and wheels mounted to the platform. The 2" hex spacers secure the outside bolts used to mount the motors. Ensure that both motors are oriented in the same direction. If they aren't, the PDA control software will have to be modified to control the direction of PDA Robot's motion. **Figure 6.23** shows the underside of the main platform with the motors, range finder, and wheels mounted. The two pieces of balsa wood under the motor gearboxes raise the

Figure 6.21

Main platform drill
diagram.

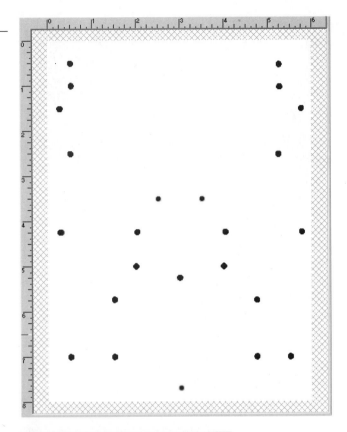

Figure 6.22

Drilled out platform
showing parts
placement.

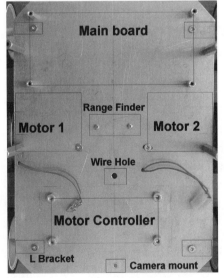

Figure 6.23

Underside of the main platform.

motors so that they are higher than the outer wheels. This ensures good traction so PDA Robot can turn easily.

Assembling the Geared Motors

I chose the Tamiya six-speed gearbox for this project and set the gear ratio to 76.5:1. This gives the craft enough power to move over dense carpet at a reasonable speed without stalling. The gear kit comes with detailed instructions on assembling the motors. **Figure 6.24** shows the step in the assembly instructions detailing the gear placement for the 76.5:1 ratio (132-rpm). **Figure 6.25** shows the assembled gearbox.

To mount the wheels on the gearboxes, insert the spring pin and use wheel hub #2 provided with the sports tire set, and fasten the wheel to the shaft using the hex wrench that comes with the kit. **Figure 6.26** shows how to mount the wheel hub on the shaft. **Figure 6.27** shows the mounted gearbox with the wheel attached.

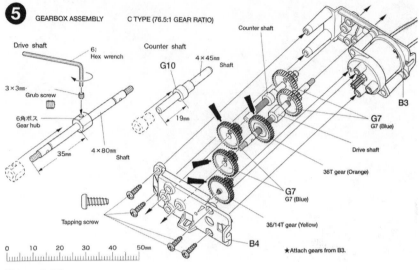

Figure 6.24

Assembling the gearbox.

Once the gearboxes have been mounted, push the motor wires through the holes and solder the Molex wire connectors to the leads. Ensure that the ground wire is inserted on the left of the plastic housing. See **Figure 6.22**.

Figure 6.25

The assembled gearbox.

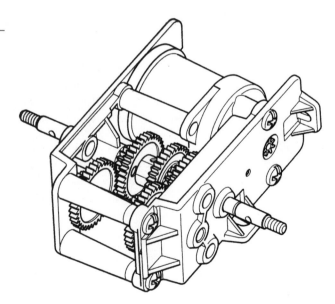

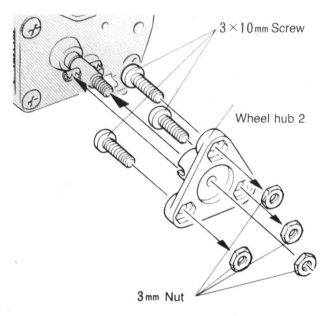

Figure 6.26

Mounting the wheel hub.

Secure the L-brackets and mount the wheels using wheel hub #1 and a 1" 4-40 bolt with a washer so that there is not too much wobble. **Figure 6.28** shows the side profile. Note: You may want to substitute the L-brackets for casters that will allow the front and back wheels to swivel freely. I found that the L-brackets work well on smooth surfaces

Figure 6.27

Mounted gearbox with wheel attached.

Figure 6.28

Side profile of PDA
Robot.

Center wheel slightly lower than outer wheels

or loose surfaces such as ceramic tile and gravel. The wheels may grab, hindering the turn ability of PDA Robot when the carpet pile is not low and tight. Another solution is to use smooth, hard plastic wheels on the front and back that don't grab.

Drill holes in the circuit board to correspond with the hex spacers attached to the main platform, and mount them with 4-40 bolts. Pass the IR range finder wire through the wire hole in the center of the platform, and insert the presoldered wire connectors into the plastic Molex housing.

The Ribbon Connectors

To connect the main board to the IR transceiver and the motor controller, we need to prepare the ribbon connectors. For the main board to motor controller connection, cut a 6" piece of ribbon six wires wide, and secure the connector to it by sliding the wire into the groves and pressing down on the top firmly until it is tight. Then slide the locking key in to hold everything together permanently. It is important that pin 1 of each connector goes to pin 1 of the other. Secure one connector to the ribbon, flip it over and connect it the same way on the other side. The red wire (wire 1) is always on the left. **Figure 6.29** shows the process of preparing the ribbon connector. Do the same for the IR transceiver. A shorter piece of cable about 4" should work.

It is important that the connectors are placed in the correct orientation or the circuit will not function. The pins of one connector must match up with the pins of the other. **Figures 6.30** to **6.33** show the connector's orientation and how the cables should be aligned. As a general rule, the red wire (wire 1) should always be over pin 1 of the connector.

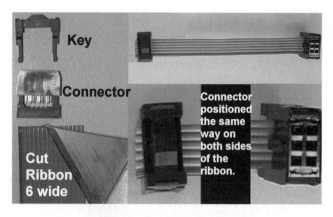

Figure 6.29

Preparing the ribbon connectors.

Attach all the connectors and drill the holes in the top plate that will support the PDA. **Figure 6.34** shows the position of the drill holes used to secure the top platform (7" × 5-1/4") to the hexagon spacers of

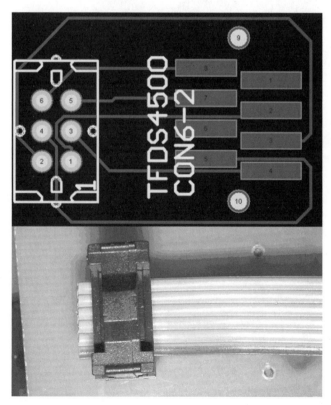

Figure 6.30

The IR transceiver connector orientation.

Figure 6:31

The IR transceiver connector orientation to main board.

Figure 6.32

The motor controller connector orientation on the motor board.

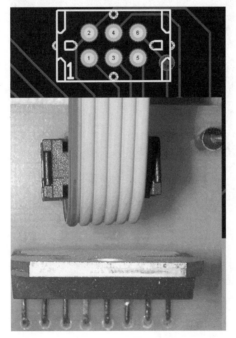

Figure 6.33

The motor controller connector orientation on the main board.

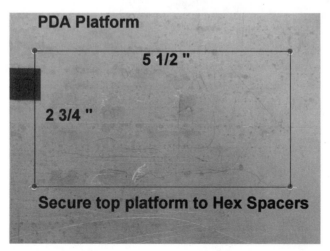

Figure 6.34

Top platform drill holes.

133

the main platform. Secure a piece of sticky Velcro to the top plate (where you would like the transceiver to go) and to the transceiver itself. We need to program the 16F876 microcontroller, so it's best to leave the top plate off until this is done (see the next chapter).

The Camera (Accessory) Mount

Drill two holes in the 1" × 1/2" × 1/4" piece of aluminum. One hole is used to secure it to the hex spacer positioned on the front of PDA Robot and the other to mount the camera. **Figure 6.35** shows the camera mount attached to the hex space. A X10 wireless video camera will be mounted here to provide vision when PDA Robot is being controlled remotely from a PC connected to the wireless network.

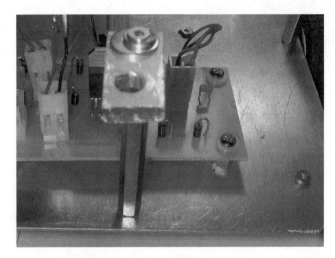

Figure 6.35

Camera mount attached to the 2" hex spacer.

Now that PDA Robot's physical body is complete, we need to give him a brain. Information on how to program the microcontroller and the PDA software is in the chapters to follow. **Figures 6.36** and **6.37** show PDA Robot fully assembled.

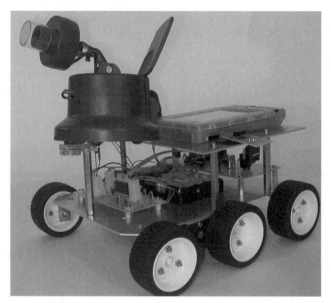

Figure 6.36

PDA Robot being controlled by a Palm OS device (Visor Deluxe).

Figure 6.37

PDA Robot being controlled with a Pocket PC device (iPAQ).

7
Programming the PIC16F876 Microcontroller

The PIC compiler is used in this project to write the software running on the PIC16F876 microcontroller, and the EPIC Plus Programmer is used to download the software to the PIC16F84A. The PIC16F876 receives input data and commands from the infrared (IR) module and the PDA via the MCP2150. It sends information such as range data and motor control confirmation codes back to the PDA. The PIC16F876 could be considered the main node of the robot's nervous system. **Figure 7.1** shows the EPIC Plus microcontroller programmer with the PIC16F876 inserted into the ZIF adapter.

The pocket-sized EPIC Plus Programmer quickly and easily programs most PICmicro microcontrollers, including the PIC16C55x, 6xx, 7xx, 84, 9xx, PIC16CE62x, PIC16F62x, 8x, 87x, PIC14Cxxx, PIC17C7xx, PIC18Cxxx, 18Fxxx, the 8-pin PIC12Cxxx, PIC12CExxx, and the 14-pin 16C505 microcontrollers. The basic programmer includes an 18-pin socket for programming 8-, 14-, and 18-pin PICmicro microcontroler unit (MCUs). (It will not program or read the baseline PIC16C5x or high-end 17C4x series.) A wide variety of adapters allow the EPIC Plus to program devices in many different package formats such as DIP, SOIC, PLCC, SSOP, TSOP, etc.

The EPIC Plus Programmer is software upgradeable for future PICs. It includes DOS and Windows 95/98/Me/NT/2000 programming software and a PIC macro assembler that works with both the Microchip

Figure 7.1

The EPIC Plus microcontroller programmer with zoomed in ZIF socket.

mnemonics and the Parallax "8051" style mnemonics. For this project, I am using the PICmicro MCU C Compiler. It can also be used with MPASM, a BASIC compiler, or a number of other C compilers.

The EPIC Plus Programmer can be powered with two 9-volt batteries or an optional AC adapter (recommended). The EPIC Plus Programmer connects to a PC compatible parallel printer port and includes an assembler and programming software. It is available either assembled or as a bare board.

Software Installation

The EPIC Plus Programmer files are compressed into a self-extracting file on the disk. They must be uncompressed before use. To uncompress the files, first create a directory on your hard drive called EPIC, or another name of your choosing by typing:

```
c:\
md epic
```

at the DOS prompt.

Change to this directory:

```
cd epic
```

Assuming the distribution diskette is in drive a:, uncompress the files into the EPIC subdirectory:

```
a:\epic2xx -d
```

where epic2xx is the name of the self-extracting file on the diskette. Don't forget the -d option on the end of the command. This ensures that the proper subdirectories within EPIC are created. Alternatively, INSTALL.BAT can be run to perform similar steps. If the EPIC directory already exists, you will get an error message and the installation will continue.

Hardware Installation

Make sure there are no PICmicro MCUs installed in the EPIC programming socket or any connected adapter's socket until the programming software is executed and the light-emitting diode (LED) is off. Also, be sure that the EPIC Plus Programmer is placed on an insulated surface to prevent the shorting out of traces on the bottom.

Connect the EPIC Plus Programmer to a PC compatible parallel printer port using a 25-pin male to 25-pin female printer extension cable. The EPIC Plus Programmer uses pins 2–6, 10, and 19–25. A serial cable may not have all of the necessary connections, so be sure to use a printer extension cable. A suitable cable is available from Micro-Engineering Labs.

Make sure the programmer is connected to a parallel printer port. Connection to a serial port or SCSI port that has similar connectors may result in damage to the port or to the programmer. If you are powering the EPIC Plus Programmer with the optional AC adapter, plug it into the power connector on the programmer and then into a wall outlet. The AC adapter should provide approximately 16VDC at 500ma.

When an AC is used adapter to power the programmer, the state of the "Batt ON" jumper does not matter.

If you are powering the EPIC Plus Programmer with two 9-volt batteries, plug each battery onto the battery snaps. Connect the 2-pin shorting jumper to the 2-pin "Batt ON" posts. It is a good idea to check the battery voltage from time to time or if there seems to be difficulty programming parts.

Warning: Do not connect a battery across the center snaps. Doing so shorts out the battery and may cause it to explode.

Note: The LED may be lit at this point. It should go out when the EPIC programming software is run. Do not insert or remove a PICmicro MCU when the LED is on.

The EPIC Plus Programmer should now be powered up and ready to program PICmicro MCUs.

General Operation

The next task is simply to write your program using any text editor, such as DOS Edit or Windows Notepad, and assemble it using the assembler, PM, included on the disk, or MPASM (or MPLAB), available from Microchip. Instructions for the use of PM are on the included disk.

Note: For PDA Robot, I am using the PICmicro MCU compiler. The source code and the process of generating the .HEX file is explained in detail in the next section of this chapter.

Once your program assembles properly, the generated .HEX file may be programmed into a PICmicro MCU using the EPIC programming software. Three versions of the EPIC software are included: two versions for DOS (one command line and one graphical) and one for Windows 95/98/ME/NT/2000/XP. If you choose the graphical DOS version, it should be used in a straight DOS session or from a full-screen DOS window in Windows 95/98 or OS/2. (Running the graphical DOS version of EPIC under Windows is discouraged. Windows [all varieties] alters the system timing and plays with the ports when you are not looking, which may cause programming errors.)

The Windows 95/98/ME/NT/2000/XP version should, of course, be run under Windows 95, 98, ME, NT, 2000, or XP. The Windows and command line DOS versions are more up to date than the graphical DOS version, and are able to program more types of PICmicro MCUs.

EPIC for DOS

Start the DOS version of the EPIC software by typing "epicdos" at the DOS command prompt in the directory you created previously. The EPIC software will look around to find where the EPIC Plus Programmer is attached, and get it ready to program a PICmicro MCU. If the EPIC Programmer is not found, check all of the above connections and verify that no PICmicro MCU is installed in the programmer or any connected adapter.

Once the programming screen is displayed, select the device type you wish to program. For PIC16C8x or PIC16F8x parts, select 8x. For the PIC14C000, PIC16C55x, 6x, 7x, or 9x parts, select 6x/7x/9x. For PIC12C5xx parts, select 12C50x.

Enter "Alt-O" (or click "Open" with the mouse) to open your assembled object (.HEX) file. Double-click on the appropriate file to load it. Once the file has been loaded, make sure the proper device characteristics are selected. See the Microchip data books for information on device configuration.

Caution: Be sure that Code Protect is set to OFF before programming a windowed (JW) PICmicro MCU. You may not be able to erase a windowed part that has been code protected.

Insert a PICmicro MCU into the EPIC Plus Programmer or connected adapter socket. The end of the PICmicro MCU with the notch should be all the way at the Pin 1 end of the socket, away from the battery connectors. Press "Alt-P" (or click "Program" with the mouse) to program the PICmicro MCU.

Before programming, the EPIC software does a blank check to ensure that the part is erased. PIC12Cxxx parts are not completely blank from the factory. They contain a calibration value in the last location. Simply tell EPIC that it is OK to program them anyway, when it finds they are not blank. If the PICmicro MCU is a 16F84 or another EEPROM or flash part, it is usually not necessary to erase it before programming.

Typing "epicdos /?" at the DOS command prompt will display a list of available options for the EPIC software.

EPIC for Windows 95/98/ME/NT/2000/XP

Because the Windows version is the simplest and most up-to-date version, I will explain how to program the PIC16F876 using it.

Start the Windows 95/98/ME/NT/2000/XP version of the EPIC software by navigating to the EPIC directory using Explorer and double-clicking on EPICWin. Alternatively, you can create a shortcut to EPIC on your desktop and double-click it. The EPIC software will look around to find where the EPIC Plus Programmer is attached and get it ready to program a PICmicro MCU. If the EPIC Programmer is not found, check all of the above connections and verify there is not a PICmicro MCU installed in the programmer or any connected adapter. The file EPIC.INI must be in the same directory EPICWIN.EXE resides in, and the EPIC directory should be in your path so that Windows can find the device drivers. Once the programming bar is displayed, select the device type you wish to program. **Figure 7.2** shows the main window with the 16F876 device selected as the target.

Click the Open button or File/Open with the mouse to open your assembled object (.HEX) file. Double-click on the appropriate file to load it. Once the file has been loaded, make sure the proper device characteristics are selected under the Options menu. See the Microchip data books for information on device configuration.

Caution: Be sure that Code Protect is set to OFF before programming a windowed (JW) PICmicro MCU. You may not be able to erase a windowed part that has been code protected.

For the PIC16F876 and the crystal oscillator used with PDA Robot, ensure that the crystal is set to High Speed (HS) and enable the Power-up timer and Brown-out reset under the Configuration menu. Use the default values for everything else in the configuration menu. **Figure 7.3** shows the settings required.

Figure 7.2

EPICWin main window.

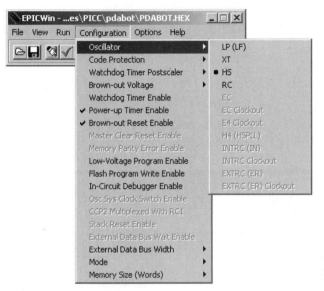

Figure 7.3

PIC16F876 configuration options.

The Options I like to have set to ensure that everything works correctly are Program/Verify Code, Program/Verify Configuration, Program/ Verify Data, Reread File Before Programming, Erase Before Programming, and Verify After Programming. Even though it is not necessary to erase the 16F876 before programming, I like to ensure that it is because I have had the odd problem when I don't erase it. **Figure 7.4** shows the Options menu.

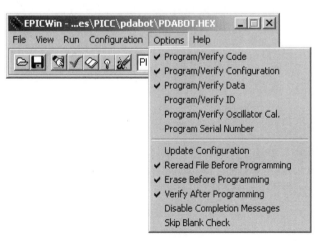

Figure 7.4

PIC programming options.

Insert a PICmicro MCU into the EPIC Plus Programmer or connected adapter socket. The end of the PICmicro MCU with the notch should be all the way at the Pin 1 end of the socket, away from the battery connectors. Click the Program button or Run/Program with the mouse to program the PICmicro MCU.

Before programming, the EPIC software does a blank check to ensure that the part is erased. If the PICmicro MCU is a 16F84 or another EEP-ROM or flash part, it is usually not necessary to erase it before programming. PIC16F7x and PIC18Fxxx devices do require erasing each time before the MCU may be reprogrammed.

The current setup is saved to the file EPICCFG.INI when you exit EPICWin. It is reloaded the next time EPICWin is started.

EPICWin Controls

The Open speed button opens a .HEX file for programming. The name of an open file appears in the EPICWin title bar. Previous configuration information will not be altered if Options/Update Configuration is not checked.

The Save speed button will save the current code, data, ID, and configuration information to the currently open file. If no file has been previously selected, it will prompt for a filename.

The Program speed button will program the current code, data, ID, and configuration into the selected device. It will optionally load the latest version of the .HEX file before programming. The device will be checked to ensure it is blank before programming, unless Options/Skip Blank Check is checked.

The Verify speed button will compare the current code, data, ID, and configuration to the programmed device. If the information does not match, an error message is displayed. A verify is also done as the device is being programmed. A code protected device cannot be verified.

The Read speed button will read the current code, data, ID, and configuration from the selected device. The configuration information will not be read if Options/Update Configuration is not checked.

The Blank Check speed button will read the code space to ensure a device is blank. It will not check the data space, ID, configuration, or the oscillator calibration word programmed by the factory into some devices.

The Erase speed button will erase EEPROM or flash electrically erasable devices. It is grayed out for devices that cannot be electrically erased.

The Device box allows selection of the device to be programmed. Click the down arrow to the right of the box to drop down a list of supported devices, then click on the device. This device information, including the default device that is selected on start-up, is contained in the file EPIC.INI. This file must be in the same directory as EPICWIN.EXE. Select the device before a .HEX file is opened to ensure the configuration information is properly interpreted. Devices with parentheses after them indicate that they will program either the base version of the part, or the version contained within the parentheses. For example, selecting the device listed as PIC16F84(A) means that either the PIC16F84 or the PIC16F84A may be programmed.

All of the speed buttons, along with other settings, are also available using the drop-down menus.

The PICmicro MCU Compiler

The code for PIC16F876 used in PDA Robot was compiled using the PICmicro MCU compiler. The code is written in C, and will be explained in detail in this chapter.

The PCM compiler is for 14-bit opcodes, and PCH is for the 16- and 18-bit PICmicro MCU. This compiler is specially designed to meet the special needs of the PICmicro MCU controllers. These tools allow developers to quickly design application software for these controllers in a highly readable high-level language.

The compilers have some limitations when compared to a more traditional C compiler. The hardware limitations make many traditional C compilers ineffective. As an example of the limitations, the compilers will not permit pointers to constant arrays. This is due to the separate code/data segments in the PICmicro MCU hardware and the inability to treat ROM areas as data. On the other hand, the compilers have knowledge about the hardware limitations and do the work of deciding how to best implement your algorithms. The compilers can implement very efficiently normal C constructs, as well as input/output operations and bit twiddling operations.

The Command Line Compiler

The command line compiler is invoked with the following command:

CCSC   options   cfilename

Valid options:

+FB	Select PCB (12-bit).	-D	Do not create debug file.
+FM	Select PCM (14-bit).	+DS	Standard .COD format debug file.
+FH	Select PCH (PIC18XXX).	+DM	.MAP format debug file.
+F7	Select PC7 (PIC17XXX).	+DC	Expanded .COD format debug file.
+FS	Select PCS (SX).	+Yx	Optimization level x (0-9).
+ES	Standard error file.	+T	Create call tree (.TRE).
+EO	Old error file format.	+A	Create stats file (.STA).
-J	Do not create PJT file.	-M	Do not create symbol file.

The xxx in the following is optional. If included it sets the file extension:

+LNxxx	Normal list file.	+O8xxx	8-bit Intel HEX output file.
+LSxxx	MPASM format list file.	+OWxxx	16-bit Intel HEX output file.
+LOxxx	Old MPASM list file.	+OBxxx	Binary output file.
+LYxxx	Symbolic list file.	-O	Do not create object file.
-L	Do not create list file.		

+P	Keep compile status window up after compile.
+Pxx	Keep status window up for xx seconds after compile.
+PN	Keep status window up only if there are no errors.
+PE	Keep status window up only if there are errors.
+Z	Keep scratch and debug files on disk after compile.
I="..."	Set include directory search path, for example: I="c:\picc\examples;c:\picc\myincludes"If no I= appears on the command line the .PJT file will be used to supply the include file paths.
#xxx="yyy"	Set a global #define for id xxx with a value of yyy, example:#debug="true"
+STDOUT	Outputs errors to STDOUT (for use with third-party editors).
+SETUP	Install CCSC into MPLAB (no compile is done).
+V	Show compiler version (no compile is done).
+Q	Show all valid devices in database (no compile is done).

If @filename appears on the CCSC command line command line, options will be read from the specified file. Parameters may appear on multiple lines in the file.

If the file CCSC.INI exists in the same directory as CCSC.EXE, then command line parameters are read from that file before they are processed on the command line. For example, to compile the source code and generate a .HEX file for PDA Robo, we would type the following from the PICC directory.

```
CCSC +FM +P C:\PROGRA~1\PICC\PDABOT.C
```

The Source Code

This section explains in detail the C language constructs used in the source code of the program running on the PIC16F876. I have offloaded most of the processing to the PDA so the code on the microcontroller is very straightforward. To quote Albert Einstein, "Make things as simple as possible, but not simpler." The software waits for a command from the PDA, signaling each motor to rotate forward, reverse, or stop and a command prompting PDA Robot to send the range data to the PDA. For example, the PDA can instruct the craft to turn by sending two commands, a "motor 1 forward" command and a "motor 2 reverse" command. This section describes an optimization to the code and includes the HEX listing that can be copied to a file and burned to the PIC16F876.

The following is the code listing for pdabot.c.

```
//              PDABOT.C
//
// Software for the PIC16F876 used to controlPDA Robot
//
// Author: Douglas H Williams
// PDA Robotics: McGraw-Hill 2003
//

#include <16f876.h>

//
// We are using a 20 MHz oscillator so set the clock accordingly
//

#use delay(clock=20000000)

//
// Set pins B0 & B1 as our RS232 port which are connected to the
// MCP2150 IrDA Protocol Stack Controller
//

#use rs232(baud=115200, xmit=PIN_B1, rcv=PIN_B0, stream=PDA)
```

```
main() {

    //
    // The value from the range finder
    //

    int range_value;

    //
    // The command sent from the PDA
    //

    char cmd;

    //
    // Set up port A as analog, pin A0 is connected
    // to the sharp GP2D12 infrared range finder
    //

    setup_port_a( ALL_ANALOG );
    setup_adc( ADC_CLOCK_INTERNAL );
    set_adc_channel( 0 );

    //
    // Set the B port pins that interface to the
    // L298 motor controller low to ensure no
    // motor movement on startup. Pins B2 & B3
    // control Motor 1 and B4 & B5 control Motor 2.
    //

    output_low(PIN_B2);
    output_low(PIN_B3);
    output_low(PIN_B4);
    output_low(PIN_B5);
    output_low(PIN_B6);
    output_low(PIN_B7);
    delay_cycles(3);

    //
    // Let the PDA know we are alive by sending some data (A space characater)
    //

    fprintf(PDA, " ");

    //
    // Loop indefinitely handling commands from the PDA
    //

    while (1){

        //
```

```
// Get the command sent from the PDA
//

cmd = fgetc(PDA);

//
//  Motor 1 commands
//

//
// Motor 1 Forward
//

if( cmd == 'a')
{
    output_high(PIN_B2);
    output_low(PIN_B3);
}

//
// Motor 1 Reverse
//

if( cmd == 'b')
{
    output_low(PIN_B2);
    output_high(PIN_B3);
}

//
// Motor 1 Stop
//

if( cmd == 'c')
{
    output_low(PIN_B2);
    output_low(PIN_B3);
}

//
// Motor 2 Forward
//

if( cmd == 'd')
{
    output_high(PIN_B4);
    output_low(PIN_B5);
}

//
// Motor 2 Reverse
//
```

```
if( cmd == 'e')
{
    output_low(PIN_B4);
    output_high(PIN_B5);
}

//
// Motor 2 Stop
//

if( cmd == 'f')
{
    output_low(PIN_B4);
    output_low(PIN_B5);
}

//
// The PDA has requested that we get the value from the
// Analog input of the Range Finder
//

if( cmd == 'g')
{

    //
    // Give some time for the clear to send. We could check the CLS
    // pin from the MCP2150 here by reading the port value of Pins
    // RB6 and RB7 can be configured as.inputs and used to monitor the
    // MCP2150's Request to send (RTS: pin 13) and Clear to send (CTS: pin 12)
    // because I have connected them on the circuit board. However, if the data is
    // lost the PDA will ask for it again. See Chapter 5, Figure 5.12: Schematic of
    // PIC16F876 connection to MCP2150
    //

    delay_ms(3);

    //
    // Read the analog value from the range finder
    //

    range_value = Read_ADC();

    //
    // Send the value to the PDA
    //

    putc(range_value);
    }
  }
}
```

This code can be optimized by having the PDA simply send a byte that represents the state of the pins. By doing this, we can replace the six "if" commands used to set the pins and the state of the motor with the line:

```
OUTPUT_B(value);
```

If we convert the number to binary, you can see that we need the first four bits, and the fifth bit can be used to represent a request for the rangefinder data. So if the value is less than 64, we know it is a motor command, and if higher, a request for the range.

```
010100 (binary)   = 20 (decimal) = Both motors moving forward
101000 (binary)   = 40  (decimal) = Both motors moving Reverse
...
000000 (binary)   = 0   (decimal) = Both motors stopped
1000000 (binary) = 64 decimal = The PDA has requested the range data.
```

The code preceding was written using Notepad and saved as pdabot.c in c:\Program files\picc\pdabot. The next step is to invoke the command line compiler. When you use the following command in a command prompt from the picc directory, the +P flag instructs the compiler to leave the compilation window displayed when it is complete. This allows you to see if any errors were detected in the code and on what line the compiler was having the problem.

```
CCSC +FM +P C:\PROGRA~1\PICC\PDABOT\PDABOT.C
```

Figure 7.5 shows the command prompt with the CCSC command line used to invoke the compiler and compiler the code to the .HEX file that we will burn onto the PIC microcontroller.

The compiler displays a window, shown in **Figure 7.6**, indicating the status of the compilation and information regarding the memory usage of the device.

The following is the compiled hex listing for the above source code that is loaded into the memory of the PIC16F876.

```
0000- 3000 008a 283f 0000 3008 00f7 1683 1406
0008- 1283 1806 2809 01af 17f7 281c 13f7 281c
0010- 1003 1806 1403 0caf 1777 281c 1377 0bf7
0018- 2810 082f 00f8 2829 3016 1bf7 3006 00f8
0020- 0bf8 2820 0000 0000 1bf7 280e 1b77 2816
0028- 2810 118a 120a 2862 302f 0084 0800 1903
0030- 283e 3006 00f8 01f7 0bf7 2834 0bf8 2833
0038- 307b 00f7 0bf7 283a 0b80 2831 3400 0184
```

Figure 7.5

Command line
compilation.

Figure 7.6

CCSC compilation
window.

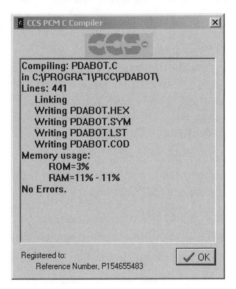

```
0040- 301f 0583 3007 1683 009f 1086 1283 1486
0048- 01a1 1683 1106 1283 1106 1683 1186 1283
0050- 1186 1683 1206 1283 1206 1683 1286 1283
0058- 1286 1683 1306 1283 1306 1683 1386 1283
0060- 1386 2804 0878 00a4 0b24 2866 1683 1106
0068- 1283 1506 3003 00af 202c 1683 1186 1283
0070- 1186 0824 3c02 1d03 2875 1683 1186 1283
0078- 1586 3003 00af 202c 1683 1106 1283 1106
0080- 0824 3c03 1d03 2884 1683 1106 1283 1106
```

```
0088- 3003 00af 202c 1683 1186 1283 1186 0824
0090- 3c72 1d03 2893 1683 1106 1283 1506 1683
0098- 1186 1283 1186 3003 00af 202c 0824 3c0f
00a0- 1d03 28ad 1683 1206 1283 1606 1683 1286
00a8- 1283 1686 3003 00af 202c 0824 3c66 1d03
00b0- 28bc 1683 1106 1283 1106 1683 1186 1283
00b8- 1586 3003 00af 202c 0824 3c73 1d03 28c0
00c0- 1683 1106 1283 1106 1683 1186 1283 1186
00c8- 3003 00af 202c 0824 3c78 1d03 28cf 0824
00d0- 3c64 1d03 28d6 3003 00af 202c 0824 3c6d
00d8- 1d03 28dd 3003 00af 202c 2861 0063
```

Program the PIC16F876

Once the pdabot.hex file has been compiled, start the EPIC win program after placing the PIC16F876 into the ZIF socket and pulling the lever down. Open pdabot.hex, set the options described above, and press the Run button on the user interface (see **Figure 7.7**).

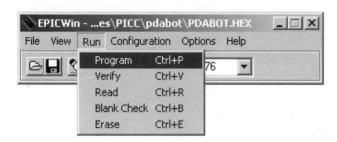

Figure 7.7

Programming the PIC16F876.

Progress windows will pop up, indicating the status of the operation being performed (see **Figure 7.8**).

When the programming is complete, a window indicating this will pop up (see **Figure 7.9**).

The PIC16F876 can now be inserted into the IC socket on the main board of PDA Robot.

Figure 7.8

Programming
status.

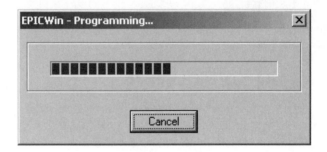

Figure 7.9

Programming
complete.

Discover | Connect | Dis

Forward

Left | Stop | Right

Reverse

Get Range

8

PDA Robot
Palm OS
Software Using
Code Warrior 8.0

I chose CodeWarrior 8.0 for this project for a number of reasons. Metrowerks provides a free evaluation copy that lets you become familiar with the intuitive integrated design environment (IDE), and everything (including the emulator) is bundled in a simple to install package. The help that it provides is excellent. Everything has compiled, linked, and worked without any problems, spectacularly. The evaluation version has some limitations, however, like a limited code size and the inability to link with additional software development kit (SDK) libraries. If you get great job writing code, buy a full-blown version.

The evaluation version can be found at the following URL: http://www.metrowerks.com/MW/Secure/Eval/Palm/default.htm. After filling out some information, the evaluation serial number to unlock the installation is e-mailed to you.

The program name is PDA Robot and the executable program that is installed on the PDA is PDARobot.prc. It creates an Infrared Data Association (IrDA) link with the robotic system, also named PDA Robot, sends, receives, and interprets commands. The code demonstrates obstacle avoidance by checking the range finder data and making a decision to turn, based on a predefined distance threshold. The standard Palm infrared (IR) library and the code supplied here are used to achieve the IrDA link with PDA Robot. The creator code "PDAr" has been registered with the Palm OS site.

This chapter includes the steps required to complete the program, while providing and explaining the C code in detail. This includes designing the graphical interface, the IrDA link, and tying it all together. **Figure 8.1** shows the main screen of PDARobot.prc.

Figure 8.1

PDA Robot.prc main screen.

CodeWarrior for Palm OS Platform version 8.0 is a fully integrated development environment for the Palm OS platform. It comes with the 4.0 SDK (the latest version of the SDK available at the time this book went to print), so you can develop for any device that runs any version of the Palm OS from the original 1.0 up to the recent 4.0. This is done through the marriage of C or C++ code and the resources that make up the user interface (UI). Resources include anything the user can interact with—forms, menus, buttons, etc. This section includes a list of the various UI elements that can be added to a Palm OS application.

First is the required copyright statement: Metrowerks and the Metrowerks logo are trademarks or registered trademarks of Metrowerks Corp. and/or its subsidiaries in the United States and other countries. CodeWarrior, PowerPlant, Metrowerks University, CodeWarrior Constructor, Geekware, and Discover programming are trademarks and/or registered trademarks of Metrowerks Corp. in the United States and other countries.

Creating the PDA Robot Project

Start the CodeWarrior IDE. Clicking on "File/New" brings up the new project dialog that allows us to select the Palm OS Application Wizard and enter the project name and where it will be saved (see **Figure 8.2**) Enter PDARobot and select a directory where you want the source code to reside (see **Figure 8.3**).

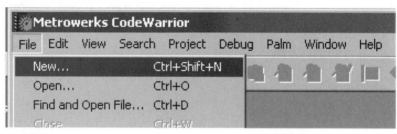

Figure 8.2

Creating a new project.

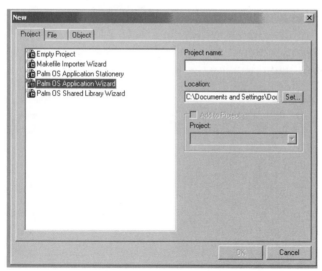

Figure 8.3

Palm OS Application Wizard dialog.

Figure 8.4

Application
information.

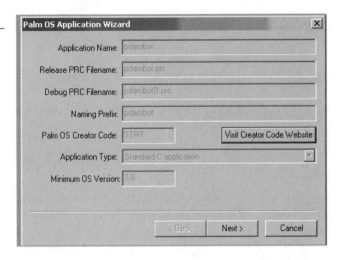

Click OK after entering the information and the dialog shown in
Figure 8.4 is displayed, showing the application information.

The additional SDKs to add are grayed out because this is the evaluation version. No additional SDKs are needed for this project anyway.
The Creator Code and Minimum OS Version is grayed out as well, but
this will be changed in the code later.

Click Finish to create the project files and open the project window
(see **Figure 8.5**).

Click on the source folder and then double click on pdarobotMain.c to
open this file for editing. First, open the Constructor for the Palm OS
1.6 by clicking on the Resources folder and double-click the file pdaro-
bot.rsrc. The constructor window will appear with the project

Figure 8.5

The PDA Robot
project window.

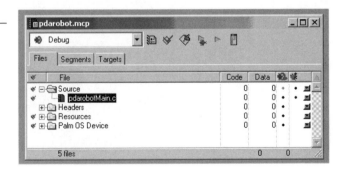

resources viewable in a dialog. From here, we will create the buttons and labels that will make the graphical user interface (GUI) for PDARobot.prc.

Double click Main under Forms to bring up the main dialog, allowing us to place the buttons and labels, after which we can assign the IDs needed for tracking in the event loop of main.c (see **Figures 8.6** and **8.7**).

Clicking Window/Catalog will bring up the catalog window that contains the controls to be placed on the form (see **Figure 8.8**).

Figure 8.9 shows the form with the buttons and labels in place. The IDs and captions have been assigned to each. I made the Object

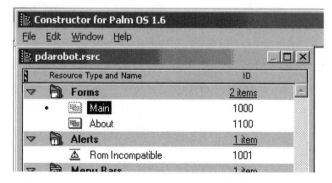

Figure 8.6

Portion of the constructor menu.

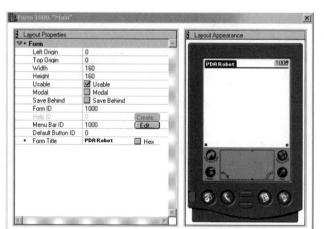

Figure 8.7

Clean palette where the controls will be placed.

159

Figure 8.8

UI Objects.

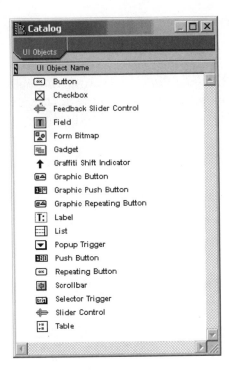

Identifier the same as the Label displayed on each Object. To generate the header file used when we compile and link the program, click File/Generate Header File or simply click File/Save.

Figure 8.9

Form with the controls placed.

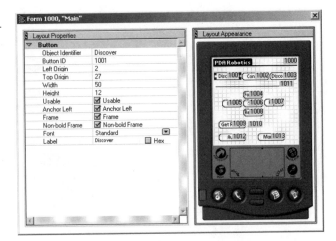

```
C:\source\pdabot\pdarobot>cd Obj

C:\source\pdabot\pdarobot\Obj>dir *.prc
 Volume in drive C is Local Disk
 Volume Serial Number is F435-A930

 Directory of C:\source\pdabot\pdarobot\Obj

22/03/2003  10:38p                 4,514 pdarobot.prc
22/03/2003  10:34p                 4,693 pdarobotD.prc
               2 File(s)            9,207 bytes
               0 Dir(s)    19,210,534,912 bytes free
```

Figure 8.10

The Release and Debug executables in the OBJ directory.

If we switch back to the Metrowerks CodeWarrior IDE and click Project/Make or hit F7, the application will build and generate PDARobot.prc that can be loaded on the PDA and run (though nothing will happen when you press the buttons).

It can be run on the Windows desktop by starting the emulator provided by Palm OS (that was installed with the evaluation version of CodeWarrior). To do this, start the emulator (after downloading or acquiring a ROM of the device) and in the IDE, click Project/Run or hit Ctrl+F5. **Figure 8.11** shows the program so far, running in the Palm OS Emulator. It looks exactly the same running on the device.

Figure 8.11

PDA Robot running on the Palm OS Emulator.

The AppStart() function reads in any saved information and initializes the infrared libray by calling the function StartApplication, which I should have called InitializeInfrared().

```
// FUNCTION: AppStart
//
// DESCRIPTION:  Get the current application's preferences.
//
// RETURNED:
//     errNone - if nothing went wrong

static Err AppStart(void)
{
    UInt16 prefsSize;

    // Read the saved preferences / saved-state information.
    prefsSize = sizeof(pdarobotPreferenceType);
    if (PrefGetAppPreferences(
        appFileCreator, appPrefID, &g_prefs, &prefsSize, true) !=
        noPreferenceFound)
    {
        // FIXME: setup g_prefs with default values
    }

    if (!StartApplication())
        return 0;

    return errNone;
}
```

The AppStop() function saves any preferences and calls StopApplication() which shuts down the infrared communication.

```
// FUNCTION: AppStop
//
// DESCRIPTION: Save the current state of the application.

static void AppStop(void)
{
    // Write the saved preferences / saved-state information.  This
    // data will be saved during a HotSync backup.
    PrefSetAppPreferences(
        appFileCreator, appPrefID, appPrefVersionNum,
        &g_prefs, sizeof(pdarobotPreferenceType), true);

    StopApplication();

    // Close all the open forms.
    FrmCloseAllForms();
}
```

StartApplication loads the IR library, opens and binds a port, saving the information in the variable irref so it can be used elsewhere. This is the *first* step in creating the IrDA link.

```
//
// Loads the Infrared Library and opens and binds the port.
//

static Boolean StartApplication(void )
{
    if (SysLibFind(irLibName,&irref) != 0)
    {
        FrmAlert(IrLibProblemAlert);
        return false;
    }
    else
    {
        if (IrOpen(irref,irOpenOptSpeed115200) != 0)
{
            FrmAlert(IrLibProblemAlert);
            return false;
        }
    }

    IrSetConTypeLMP(&connect);
    packet.buff = (unsigned char *)"Data";
    packet.len = 4;

    IrBind(irref,&connect,callback);

    return true;
}
```

StopApplication unbinds the port, disconnects, and closes the IR connection.

```
//
// Shut down connections, close the library
//

static void StopApplication(void)
{
    IrUnbind(irref,&connect);

    if (IrIsIrLapConnected(irref))
        IrDisconnectIrLap(irref);
    IrClose(irref);
}
```

The callback(IrConnect *con, IrCallBackParms *parms) function is called whenever an infrared event happens, for example, when PDA Robot sends us some data, this function is automatically called with the event and data embedded in the parms parameter.

```
static void callback(IrConnect *con, IrCallBackParms *parms)
{
    char* event;

    char out= 0;
    FormType *frm; // = FrmGetActiveForm();

    switch (parms->event)
    {
    case LEVENT_DISCOVERY_CNF:

        //
        // This event was triggered by PDA Robot when
        // we broadcast a discovery to ALL IrDA compliant
        // devices. StoreDiscovery throws away all devices
        // except PDA Robot. My HP printer always responds to
        // the discovery request.
        //

        event = "DISCOVERY_CNF";
        StoreDiscovery(parms->deviceList); break;


    case LEVENT_PACKET_HANDLED:
        packet_in_use = false;
        event = "PACKET_HANDLED"; break;

    case LEVENT_DATA_IND:

        //
        // PDA Robot has sent some data  because we requested it.
        // Let's copy the data to a global variable so it can be used
        // elsewhere.
        //

        event = "DATA_IND";
        MemMove(&received_data, parms->rxBuff, parms->rxLen);

        MemMove(&out, &received_data[1], 1);
        frm = FrmGetActiveForm ();
        FrmDrawForm(frm);

        StrPrintF((char *) range_data, "%u", out);//value);
        current_range = out;

        //
```

```
// Display the range in the Range Label if we are in autonomous mode
//

if( autonomous == true )
{
    FrmCopyLabel (frm, MainRangeLabel, (char*)&range_data);
}

range_aquired = true;
FrmDrawForm(FrmGetActiveForm());
break;

case LEVENT_STATUS_IND:
    switch (parms->status)
    {
        case IR_STATUS_NO_PROGRESS:
            event = "S_NO_PROGRESS"; break;
        case IR_STATUS_LINK_OK:
            event = "S_LINK_OK"; break;
        case IR_STATUS_MEDIA_NOT_BUSY:
            event = "S_MEDIA_NOT_BUSY";        break;
        default:
            event = "S_UNKNOWN";
    }
    break;
case LEVENT_TEST_CNF:
    switch (parms->status)
    {
        case IR_STATUS_SUCCESS:
            event = "TEST_SUCCESS"; break;
        case IR_STATUS_FAILED:
            event = "TEST_FAILED"; break;
    }
    break;
case LEVENT_TEST_IND:
    event = "TEST_IND"; break;
default: event = "UNKNOWN";
}

}

//
// StoreDiscovery goes through the devices list returned when we
// sent out a Discovery request to all IrDA devices in the vicinity.
// It throws away all devices except PDA Robot and set the connection
// information returned to us by it.
//

void StoreDiscovery(IrDeviceList* deviceList)
{
    UInt8 i;
```

```
char info[36];

// clear the label
StrCopy((char *)&info, (char *)"_____");
FrmCopyLabel (FrmGetActiveForm(), MainStatusLabel, (char*)&info);

if(  deviceList->nItems == 0 )
{
    StrCopy((char *)&info, (char *)"NO Devices Discovered ");
    FrmCopyLabel (FrmGetActiveForm(), MainStatusLabel, (char*)&info);
    return;
}

for (i = 0; i < deviceList->nItems; i++)
{
    //
    // We don't want to recognize any device but PDA Robot
    // so ensure that the device name is 'Generic IrDA'. This
    // is the default name used by the MCP2150 chip. We will
    // connect with the first found
    //

    if( (StrCompare((char *)"Generic IrDA", (char *) &deviceList->dev[i].xid[3])) == 0)
    {
        dev = deviceList->dev[i].hDevice;
        connect.rLsap = deviceList->dev[i].xid[0];

        StrCopy((char *)&info, (char *)"Discovered PDA Robot   ");
        FrmCopyLabel (FrmGetActiveForm(), MainStatusLabel, (char*)&info);
    }
}

}

//
// Information Access Service Callback. This function
// is called when we query PDA Robot for information.
// If we received the LSAP information then we connect
// to to PDA Robot.
//

static void IrlasCallback(IrStatus status) {

    UInt8 b;
    UInt8 i;

    if((query.retCode)!=IAS_RET_SUCCESS)
    {
        return;
    }

i=IrIAS_GetType(&query);
```

```
switch(i)
{
case IAS_ATTRIB_MISSING:
    break;

case IAS_ATTRIB_INTEGER:

        if(rtype!=0)
    {
            rlsap = connect.rLsap = IrIAS_GetIntLsap(&query);
        }

        connect.rLsap = rlsap;
        packet.buff = (unsigned char *)&controlPacket;
        packet.len = sizeof(controlPacket);

    //
        // Open a connection with PDA Robot
        //

        IrConnectReq(irref, &connect, &packet, DEFAULT_TTP_CREDIT);
        rtype=0;
        break;

case IAS_ATTRIB_USER_STRING:
        b=IrIAS_GetUserStringCharSet(&query);
FrmCopyLabel (FrmGetActiveForm (), MainRangeLabel,
(char*)IrIAS_GetUserString(&query));
    break;

default:
        //
        // Unknown IAS Reply
        //
    break;
    }

}
```

Please go to www.pda-robotics.com to download the entire source code and executable for this program.

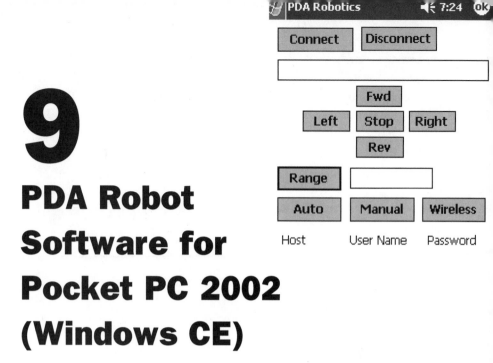

9

PDA Robot Software for Pocket PC 2002 (Windows CE)

The software for the Pocket PC was designed and written using the eMbedded Visual Tools 3.0 IDE and compiler, in conjunction with the Pocket PC 2002 Software Development Kit (SDK). Both are provided free from Microsoft.

The Microsoft eMbedded Visual Tools 3.0 deliver a complete desktop development environment for creating applications and system components for Windows-powered devices, including the Pocket PC and Handheld PC.

The eMbedded Visual Tools include eMbedded Visual Basic and eMbedded Visual C++, including SDKs for the Pocket PC 2000, Palm-size PC, and Handheld PC. The eMbedded Visual Tools are the successor to the separate Windows CE Toolkits for VC++ and VB. This version is stand-alone and does not require Visual Studio. Read the specifications on the data sheet.

The Pocket PC 2002 SDK allows you to write enterprise and consumer applications for this innovative platform. The Pocket PC 2002 SDK provides a brand new Pocket PC 2002 device emulator, more documentation, and more samples. In addition, this SDK includes all the necessary application programming interfaces (APIs) and documentation for both Pocket PC 2002 and Pocket PC 2002 Phone Edition devices.

Install eMbedded Visual Tools 3.0 (eVC 3.0) first, followed by Pocket PC 2002. The Pocket PC 2002 installation will point the eVC 3.0 to the correct header and library files, install the emulator, and set up the compilation targets.

Figure 9.1 shows the main window of eMbedded Visual Tools 3.0.

Figure 9.1

Embedded Visual C++ IDE.

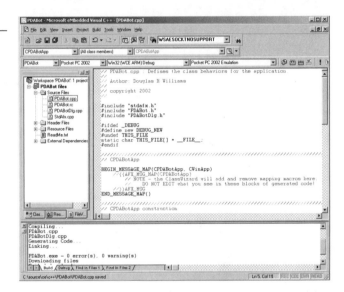

When the code is complied and run with the Pocket PC 2002 emulation option, the emulator is invoked and the program can be run on your desktop. **Figures 9.2** and **9.3** show the default emulator that ships with Pocket PC 2002 and PDABot.exe running on the emulator.

Microsoft eMbedded Visual C++ 3.0 Overview

Microsoft eMbedded Visual C++ 3.0 is the most powerful way for developers to build applications for the next generation of Windows CE-based communication, entertainment, and information-access devices. This stand-alone IDE brings a new level of productivity to Windows CE development without compromising flexibility, performance, or control.

With eMbedded Visual C++, developers can accomplish the following:

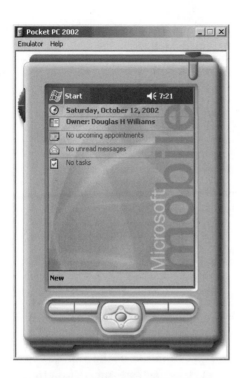

Figure 9.2

Standard Pocket PC emulator.

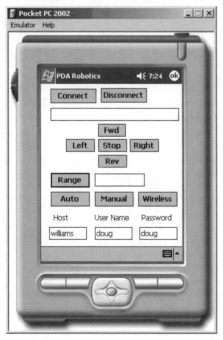

Figure 9.3

PDA Robot running on the Pocket PC emulator.

- Take advantage of a familiar development environment by building Windows CE applications using a stand-alone IDE designed to target Windows CE development;

- Access Windows CE-specific documentation targeted for the platform SDKs users have installed on their workstation;

- Save time and money by using the Windows CE version of the Microsoft Foundation Classes (MFC) and the Active Template Library; and

- Build enterprise solutions with data-access capabilities through ActiveX Data Objects (ADO) for Windows CE, transactional processing via Microsoft Transaction Server, and more through close integration with Windows CE operating system services.

Benefits of using the eMbedded Visual C++ include the following:

- Gain direct access to the features of the underlying operating system without the need for additional code, providing full control over device hardware and operating system services.

- Access all the features of every permutation of the Windows CE operating system to construct the fastest, most functional Windows CE applications.

- Be the first to program the newest and most exciting Windows CE devices, using Windows CE platform SDKs for eMbedded Visual C++ as they become available.

- Get in on the ground floor and perhaps build the "killer application" for a next-generation operating system.

- Expand development options to include a whole new group of computer users and equip those normally resistant to computers with the simplicity of Windows CE running focused applications, such as Internet browsing, task-specific business processes, or entertainment programs.

- Build highly mobile applications that can access remote data stores and communicate with networked servers.

Increased Developer Productivity

- Leverage existing knowledge and training by building Windows CE solutions from within the same development environment as that used for traditional Windows development.

- Gain increased programmer productivity with IntelliSense technology, providing on-the-fly programming assistance including statement completion, parameter information, and syntax error checking. Minimize software development effort by building reusable ActiveX components—usable from both eMbedded Visual C++ and eMbedded Visual Basic Windows CE applications.

- Quickly select and configure project deployment information within the environment to target the widest variety of Windows CE devices and processors.

Simplified Debugging and Deployment

- Quickly test and execute applications by allowing the eMbedded Visual Tools to automatically copy and launch applications on a mobile device or emulator after compilation.

- Fix bugs fast with an integrated debugger that helps eliminate errors in applications as they are running on Windows CE devices or within an emulator.

- Gain maximum control over Windows CE development through a variety of additional tools designed to provide details on application execution.

- Avoid the need for costly hardware investments by first testing applications on a Windows CE device emulator, providing the look and feel of a physical device from within a PC environment. And, with the Pluggable SDK model, new emulators can be easily added to the Toolkit as they become available.

Comprehensive Access to the Windows CE Platform

- Gain control over communication mechanisms, such as TCP/IP, running via an infrared port or serial port to build compelling mobile applications.

- Take advantage of COM, the world's most successful and powerful component model, to build reusable solutions for Windows CE-based devices.

- Maximize development effort by reusing existing ActiveX controls created for the Windows CE platform. Harness the full power of Windows CE by using eMbedded Visual C++ to access every API on all Windows CE devices.

- Graphically build applications using the CommandBar and MenuBar controls, unique Windows CE graphical elements that combine toolbars and menus onto a single control for Windows CE platforms. Allow the Platform Manager to automatically configure a connected Windows CE device for application testing and execution.

- Build compact and efficient COM servers using the Windows CE version of the Active Template Library.

- Use MFC for Windows CE, a proven application framework to build solutions for Windows CE devices, including applications using the DOC/View architecture.

Build for the Latest Windows CE Devices

- Build solutions for the Handheld PC Pro, Palm-size PC, and Pocket PC Windows CE devices with maximum mobility and minimum maintenance and administration.

- Build powerful data-retrieval and analysis applications for the Pocket PC.

- Gain maximum flexibility and quickly add to the capabilities of the Windows CE Toolkits by being able to plug in the development kits for the latest Windows CE platforms.

Fast, Flexible Data Access

- Use a subset of the powerful ADO data access mechanism found on desktop and workstation computers to build high-performance data-aware solutions.

- With Windows CE Services, maintain local copies of database tables from any data store, including Microsoft Access and SQL

Server, that are automatically synchronized upon connection to the remote data source.

Building the PDA Robot Pocket PC Application

To build the application, first, download and install the free eVC 3.0 IDE and Pocket PC 2002 SDK installations from the Microsoft site. Start the eVC IDE and click File/New. Select WCE Pocket PC 2002 MCF App Wizard and name the project. Check the central processing units (CPUs) you would like to include as targets. To use the emulator, you must check the WCE x86. Check the other CPUs such as WCE ARM.PDABot. Click OK and select Dialog Based and the language you want. Click Next and be sure to check the Windows Sockets option. Enter the title you want in the case "PDA Robotics." Click Next, Next, Finish. The project has now been created and we can begin placing the buttons and edit boxes on the screen.

Switch to the Resources tab, Under PDABot resources/Dialog, double-click on IDD_PDABOT_DIALOG. The blank form will appear, on which we will add the buttons and two edit boxes. Place the controls as shown in **Figure 9.4**.

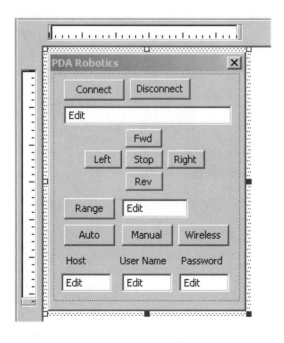

Figure 9.4

Editing the resources.

To assign an ID to a control, highlight it and hit enter. Below is the message map showing the IDs and their associated functions. When a user clicks on a button, the associated function is called. This is all handled by the Windows subsystem.

```
BEGIN_MESSAGE_MAP(CPDABotDlg, CDialog)
    //{{AFX_MSG_MAP(CPDABotDlg)
    ON_BN_CLICKED(IDC_CONNECT_IRDA, OnConnectIrda)
    ON_BN_CLICKED(IDC_CLOSE_IRDA, OnCloseIrda)
    ON_BN_CLICKED(IDC_ROBOT_FWD, OnRobotFwd)
    ON_BN_CLICKED(IDC_ROBOT_LEFT, OnRobotLeft)
    ON_BN_CLICKED(IDC_ROBOT_STOP, OnRobotStop)
    ON_BN_CLICKED(IDC_ROBOT_RIGHT, OnRobotRight)
    ON_BN_CLICKED(IDC_ROBOT_REV, OnRobotRev)
    ON_BN_CLICKED(IDC_RANGE, OnRange)
    ON_BN_CLICKED(IDC_AUTO, OnAuto)
    ON_BN_CLICKED(IDC_MANUAL, OnManual)
    ON_BN_CLICKED(IDC_WIRELESS, OnWireless)
    ON_WM_TIMER()
    //}}AFX_MSG_MAP
END_MESSAGE_MAP()
```

In order to have access to some of the objects like the edit boxes, we need to assign member variables to them. Do this by clicking View/ClassWizard and switching to the Member Variables tab. **Figure 9.5** shows the ClassWizard with the Member Variables assigned to the edit boxes. We can now get and set the information of these edit boxes because the Member Variable in the dialog gives users access to the CEdit class.

Figure 9.5

ClassWizard.

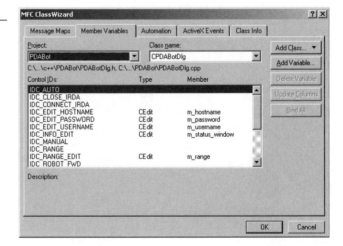

Double-clicking on a button will create a function that is called when it is clicked. Users will be directed into the function where they can add the code. The name of the function is generated by the IDE. For example, the function generated for the Wireless button is named OnWireless.

```
void CPDABotDlg::OnWireless()
{
// Add Code here
}
```

Creating the IrDA Link

The following code is used to establish the Infrared Data Association (IrDA) data link with PDABot's body. This socket could be equivalent to the information that flows up and down the spinal column, transferring information to the brain (PDA), where it makes a decision and sends a return control command. The PDA initiates a conversion with the body and asks, "Who are you?" The MCP2150 will identify itself as "Generic IrDA" until its identifier is reprogrammed to that which the designer chooses. How to do this is discussed in detail in the chapter about the MCP2150. For the explanation of this code, I will use the default MCP2150 identifier ("Generic IrDA"). The fact that a device must identify itself by the IrDA standard allows users to use this security feature by means of a keyed handshake or simply by agreeing to only accept the socket agreed upon association name.

When users click on the Connect button program control, they are directed to the function OnConnectIrda() shown in the following code. It checks to see if a link is already established by referencing the member variable m_bIrDAConnected, which is set by the return value of InitiateIrDAConnection(). I wrote the function InitiateIrDAConnection() to do the work of creating the socket and establishing the link with PDA Robot.

```
void CPDABotDlg::OnConnectIrda()
{
    //
    // Create the IrDA association with PDA Robot
    //

    if( !m_bIrDAConnected)
    {
        m_bIrDAConnected = InitiateIrDAConnection();
```

```
    }

    //
    // Disable the wireless button since we MUST first  be connetd to the command cen-
ter
    // before initializing the IrDA connection if we want to use the wireless link
    //

    m_wireless_button.EnableWindow(FALSE);

}
```

Below is the code listing for InitiateIrDAConnection().

```
bool CPDABotDlg::InitiateIrDAConnection()
{
    //
    // Initiate an IrDA client
    //

#define DEVICE_LIST_LEN 5
#define IAS_QUERY_ATTRIB_MAX_LEN 32

    //
    // DevListBuff discovery buffer stores the information that PDARobots body will send to
    // us in the initial stages of the IrDA handshake
    //

    BYTE  DevListBuff[sizeof(DEVICELIST) - sizeof(IRDA_DEVICE_INFO) +
        (sizeof(IRDA_DEVICE_INFO) * DEVICE_LIST_LEN)];
    int        DevListLen = sizeof(DevListBuff);

    //
    // This list stores all the devices that responded to our IrDA query. There may
    // be an IrDA compliant printer, like my HP1000, and the PDABot body. We
    // should look for 'Generic IrDA' and connect with only this device. I will
    // leave this modification up to you. See the chapter on the PalmOS software
    // for instructions on how to do this. For now I pick the first device in the list.
    //

    PDEVICELIST  pDevList    = (PDEVICELIST) &DevListBuff;

    //
    // buffer for IAS query
    //

    BYTE          IASQueryBuff[sizeof(IAS_QUERY) - 3 + IAS_QUERY_ATTRIB_MAX_LEN];
    int         IASQueryLen = sizeof(IASQueryBuff);
    PIAS_QUERY    pIASQuery  = (PIAS_QUERY) &IASQueryBuff;

    //
    // for searching through peers IAS response
```

```
//

BOOL        Found = FALSE;
UCHAR       *pPI, *pPL, *pPV;

//
// for the setsockopt call to enbale 9 wire IrCOMM
//

int  Enable9WireMode  = 1;

CString msg;

SOCKADDR_IRDA DstAddrIR = { AF_IRDA, 0, 0, 0, 0, "IrDA:IrCOMM" };

//
// Create the Infrared Socket
//

if ((Infrared_Socket = socket(AF_IRDA, SOCK_STREAM, NULL)) == INVALID_SOCKE{
    //
    // Get the error and display it in the status edit box
    //

    int last_error = WSAGetLastError();

    if (last_error == WSAESOCKTNOSUPPORT)
    {
        //
        // MessageId: WSAESOCKTNOSUPPORT
        //
        // MessageText:
        //
        // The support for the specified socket type does not exist
        // in this address family.
        //

        char err_buff[10];
        _itoa(last_error, &err_buff[0], 10);

        msg = "Error: ";
        msg += err_buff;
        msg = "no support for type in this address family";

        AfxMessageBox(msg);

    }else{
        msg = "Couldn't get socket ";
        this->m_status_window.SetWindowText( (LPCTSTR) msg);
    }

    return false;
```

```
}

//
// search for the peer device, In this case PDA Robot
//

pDevList->numDevice = 0;
if (getsockopt(Infrared_Socket, SOL_IRLMP, IRLMP_ENUMDEVICES, (CHAR *) pDevList,
   &DevListLen) == SOCKET_ERROR)
{
    msg = "No Peer conection";
    this->m_status_window.SetWindowText( (LPCTSTR) msg);
    return false;

}else{

    //
    // print number and name of devices found
    //

    char bu[20];

    _ultoa( pDevList->numDevice , bu, 10 );

    msg = "Num devices: ";
    msg += bu;
    msg += " Name ";
    msg += pDevList->Device->irdaDeviceName;

    this->m_status_window.SetWindowText( (LPCTSTR) msg);
}

if (pDevList->numDevice == 0)
{
    msg = "No IrDA device found";
    this->m_status_window.SetWindowText( (LPCTSTR) msg);
    return false;
}

//
// Assume first device, we should check the name of the device
// to ensure that it is 'Generic IrDA', the default name provided by the
// MCP2150 IrDA chip used on the PDA Robot circuit.
//

memcpy(&DstAddrIR.irdaDeviceID[0], &pDevList->Device[0].irdaDeviceID[0], 4);

//
// query the peer to check for 9wire IrCOMM support
//
```

```
memcpy(&pIASQuery->irdaDeviceID[0], &pDevList->Device[0].irdaDeviceID[0], 4);

//
// IrCOMM IAS attributes. see chapter on the IrDA protocol
//

memcpy(&pIASQuery->irdaClassName[0],  "IrDA:IrCOMM", 12);
memcpy(&pIASQuery->irdaAttribName[0], "Parameters",  11);


if (getsockopt(Infrared_Socket, SOL_IRLMP, IRLMP_IAS_QUERY,  (char *) pIASQuery,
    &IASQueryLen) == SOCKET_ERROR)
{
    this->m_status_window.SetWindowText( (LPCTSTR) CString("Couldn't get Ir socket
        options"));
    return false;
}

if (pIASQuery->irdaAttribType != IAS_ATTRIB_OCTETSEQ)
{
    //
    // peer's IAS database entry for IrCOMM is bad
    //

    this->m_status_window.SetWindowText( (LPCTSTR) CString("IAS database entry is
        corrupt"));
}

if (pIASQuery->irdaAttribute.irdaAttribOctetSeq.Len < 3)
{
    //
    // peer's IAS database entry for IrCOMM is bad
    //

    this->m_status_window.SetWindowText( (LPCTSTR) CString("IAS database entry is
        corrupt"));
}

//
// search for the PI value 0x00 and check for 9 wire support, see IrCOMM spec.
//

pPI = pIASQuery->irdaAttribute.irdaAttribOctetSeq.OctetSeq;
pPL = pPI + 1;
pPV = pPI + 2;

while (1)
{
    if (*pPI == 0 && (*pPV & 0x04))
    {
        //
        // It's good, don't need to check any futher
```

```
    //
    Found = TRUE;
    break;
}

if (pPL + *pPL >= pIASQuery->irdaAttribute.irdaAttribOctetSeq.OctetSeq +
    pIASQuery->irdaAttribute.irdaAttribOctetSeq.Len)
{
    break;
}

pPI = pPL + *pPL;
pPL = pPI + 1;
pPV = pPI + 2;
}

if (! Found)
{
    //
    // Peer doesn't support 9 wire mode.
    //
    msg = "peer doesn't support 9 wire mode";
    this->m_status_window.SetWindowText( (LPCTSTR) msg);
    return false;
}

//
// enable 9wire mode before we call connect()
//

if (setsockopt(Infrared_Socket, SOL_IRLMP, IRLMP_9WIRE_MODE, (const char *)
    &Enable9WireMode,
    sizeof(int)) == SOCKET_ERROR)
{
    msg = "Couldn't set socket options";
    this->m_status_window.SetWindowText( (LPCTSTR) msg);
    return false;
}

//
// Nothing special for IrCOMM from now on, we treat it as
// a normal socket. Try to connect with PDA Robot
//

if (connect(Infrared_Socket, (const struct sockaddr *) &DstAddrIR,
    sizeof(SOCKADDR_IRDA))
    == SOCKET_ERROR)
{
    msg = "Couldn't connect via IrDA";
    this->m_status_window.SetWindowText( (LPCTSTR) msg);
    return false;
```

```
}

//
// Test the connection to make sure all is good. If not
// then display an error
//

char err_buff[10];
int ret = send( Infrared_Socket, (const char *) "o\n",3, MSG_DONTROUTE);

if ( ret == SOCKET_ERROR)
{
    int last_error = WSAGetLastError();
    _itoa(last_error, &err_buff[0], 10);
    msg = "Send to socket errror error ";
    msg += err_buff;
    this->m_status_window.SetWindowText( (LPCTSTR) msg);
    return false;
}

    return true;
}
```

Once the connection has been established, users can now send commands to PDA Robot to instruct it to send range data or motion the motors. The following is the code to send a command to PDA Robot and to request the range data. Recall from the chapter on programming the PIC Microcontroller that a signals the electronics to move Motor1 forward. b – Motor1 Reverse. c – Motor1 Stop. d – Motor2 forward. e – Motor2 Reverse. f – Motor2 Stop. g – request for PDA Robot to send the range finder data. The range finder sends a value between 0 and 128, representing the distance to the front of the craft. 0 is approximately 90 cm and 128 is 10 cm from the range finder.

```
void CPDABotDlg::OnRobotFwd()
{
    char err_buff[10];

    CString msg = "Forward";

    //
    // Send the command to PDA Robot
    //

    int ret = send( Infrared_Socket, (const char *) "be", 2, MSG_DONTROUTE);

    if ( ret == SOCKET_ERROR)
    {
        //
```

```
    // Display the error in the status indicator
    //

    int last_error = WSAGetLastError();
    _itoa(last_error, &err_buff[0], 10);

    msg = "socket error";
    msg += err_buff;
    this->m_status_window.SetWindowText( (LPCTSTR) msg);
    return;
  }

  //
  // Set the status inidcator that we are moving forward
  //

  this->m_status_window.SetWindowText( (LPCTSTR) msg);
}


void CPDABotDlg::OnRange()
{

  // Below is how you would query for the range data

  char err_buff[10];
  char irda_buffer[128];
  u_long numbytes;

   int ret;

  //
  // Send PDA Robot the command prompting it to get the range data and
  // forward it to us
  //

  ret = send( Infrared_Socket, (const char *) "d", 1 , MSG_DONTROUTE);

  if ( ret == SOCKET_ERROR)
  {
      int last_error = WSAGetLastError();
      _itoa(last_error, &err_buff[0], 10);
      return;
  }

  //
  // You may want to get this data in the timer after giving PDA Robot some time to
     respond
  //

  //
  // Ensure that we won't be blocked waiting here on the function
```

```
// to read the data by calling ioctlsocket. This will indicate how much data
// is in the buffer as well.
//

    ret = ioctlsocket (Infrared_Socket, FIONREAD, &numbytes);
    if( (ret == 0) && (numbytes > 0) )
    {
        //
        // Receive what is in the buffer and set the
        // range edit box
        //
        ret = recv ( Infrared_Socket, &irda_buffer[0], 26, 0);
        this->m_range.SetWindowText( (LPCTSTR) CString(irda_buffer));
    }
}
```

To close the IrDA link, press the Disconnect button and the following function is called. It, in turn, calls CloseIrdaSocket listed below.

```
void CPDABotDlg::OnCloseIrda()
{
    CloseIrdaSocket();
}

void CPDABotDlg::CloseIrdaSocket()
{

    //
    // Purge the receive buffer and close the Socket to disconnect.
    //

    char irda_buffer[128];
    int ret;
    u_long numbytes;

    //
    // Ensure that we won't be blocked waiting here on the function
    // to read the data by calling ioctlsocket. This will indicate how much data
    // is in the buffer as well.
    //

    ret = ioctlsocket (Infrared_Socket, FIONREAD, &numbytes);
    if( (ret == 0) && (numbytes > 0) )
    {
        ret = recv ( Infrared_Socket, &irda_buffer[0], numbytes, 0);
    }

    ret = closesocket(Infrared_Socket);

    //
    // Set the member variable of this class that we use to   determine our status
    // of the link
```

```
//

    m_bIrDAConnected = false;

}
```

I have left the autonomous roaming mode code up to you. See the previous chapter on the Palm OS software for an idea of how to implement this AI-like functionality. To see how I implemented this, please visit www.pda-robotics.com to download the entire project (includes all the source code).

```
void CPDABotDlg::OnAuto()
{
    // TODO: See the chapter on PalmOS autonomous mode
    // and implement something similar. I want to leave
    // something for you to do. see www.pda-robotics to
    // download the entire project to see my implementation

}

void CPDABotDlg::OnManual()
{
    //  Disengage the Auto Mode.
}
```

The Wireless RF Link

The command center application (described in the next chapter) is the host application to which we will connect. It displays the video data to the user and allows the sending of commands to this program. The commands are interpreted and forwarded to the robot body using Infrared_Socket. The link is established using the class listed below. It is derived from the CceSocket and is a member of the CPDABotDlg class. I am using a Linksys WPC11 version 3.0 wireless PC card on my 3850 iPAQ handheld and a PC connected to a wireless digital subscriber line (DSL) router (see **Figure 9.6**). The WPC11 features the following:

- 11 Mb/ps high-speed data transfer rate compatible with virtually all major network operating systems.

- Plug-and-play operation providing easy setup.

- Full compliance with IEEE 802.11b standard high-speed data rate of up to 11 Mb/ps.

Figure 9.6

Wireless card.

PDASocket.hpp

```
//
// The class definition
//

class CPDASocket : public CCeSocket
{
    DECLARE_DYNAMIC(CPDASocket);
public:

    //
    // Constructor
    //

    CPDASocket(PURPOSE_E iPurpose=FOR_DATA);

protected:

    //
    // Called when data arrives over the wireless link
    //

    virtual void OnReceive(int nErrorCode);
};
```

PDASocket.cpp

```
//
// CPDASocket Derived from CceSocket Implementation
//

#include "stdafx.h"

#ifdef _DEBUG
```

```
#undef THIS_FILE
static char BASED_CODE THIS_FILE[] = __FILE__;
#endif

IMPLEMENT_DYNAMIC(CPDASocket, CSocket)

CPDASocket::CPDASocket(PURPOSE_E iPurpose):
    CCeSocket(iPurpose)
{

}

void CPDASocket::OnReceive(int nErrorCode)
{
    //
    // Call the ReadPDAData() that exists in
    // the CPDABotDlg class
    //
    ((CPDABotDlg *)AfxGetApp())->ReadPDAData();
    CSocket::OnReceive(nErrorCode);
}
```

CPDASocket inherits everything from CceSocket, meaning users call and access all the public member functions and variables. The virtual function OnReceive(int nErrorCode) is overridden so that users can implement their own version, but still use the underlying code and features. Note that the default socket type is set to data.

CCeSocket::CCeSocket

This constructor creates an instance of a socket object.

```
CCeSocket ( PURPOSE_E iPurpose = FOR_DATA);
```

Parameters

iPurpose specifies the enumerated constant that designates whether the socket is to be a listening socket or a data socket. It is one of the following values:

- FOR_LISTENING

- FOR_DATA

Remarks

When constructing a CCeSocket object, specify whether it is a listening socket or a data socket. After construction, call the Create method.

If you do not specify the purpose of the socket, the constructor constructs a data socket by default.

OnWireless: Implementing the CPDASocket Class

The following code is from CPDABotDlg and gets called when the user clicks the Wireless button. It creates the socket, identifying itself as the name of the PDA it is running on by calling gethostname() and then initiates the connection with the command center. If it went well, the command center will send back the message "SUCCESS." We then listen for other commands such as FORWARD, REVERSE, RIGHT, LEFT, STOP, and RANGE. The PDA sends the corresponding commands to PDA Robot via the infrared socket.

```
//
// OnWireless connects to the command center over the wireless network. NOTE: YOU
// MUST connect to the command center before initializing the IrDA. If you initialize the
// IrDA first this button will be disable until the application is restarted. This will be fixed
// in the next version which can be downloaded at www.pda-robotics.com
//

void CPDABotDlg::OnWireless()
{
    //
    // Listen on the wireless socket for commands from the
    // command center and forward them to PDA Robot on the
    // Infrared socket.
    //

    UpdateData(TRUE);

    m_hostname.GetWindowText(m_strServer);
    m_username.GetWindowText(m_strUsername);
    m_password.GetWindowText(m_strPassword);

    CheckForAuthentication();

    ::SetTimer(this->CWnd::m_hWnd, 1, 1000, NULL);
}

bool CPDABotDlg::CheckForAuthentication()
{
    if(!StartApplication())
    {
        return FALSE;
    }

    m_bClientConnected=true;
```

189

```
    char szHostName[25];

    //
    // Get the name of the PDA this is running and send it
    // to the command Centre.
    //

    gethostname(szHostName,25);
    m_pSocket->Send (szHostName,25,0);

    char szUsername[255];
    char szPassword[255];

    strcopy(szUsername,m_strUsername);
    strcopy(szPassword,m_strPassword);

    //
    //send the user name and the password
    //

    m_pSocket->Send (szUsername,255,0);
    m_pSocket->Send (szPassword,255,0);

    return TRUE;
}


//
// ConnectSocket Creates the CPDASocket which is derived from a CCeSocket
// and attempte to connect to the remote host that the control center is
// running on.
//

BOOL CPDABotDlg::ConnectSocket(LPCTSTR lpszHandle, LPCTSTR lpszAddress, UINT
nPort)
{
    m_pSocket = new CPDASocket(CCeSocket::FOR_DATA);

    if (!m_pSocket->Create())
    {
        delete m_pSocket;
        m_pSocket = NULL;
        this->m_status_window.SetWindowText( (LPCTSTR) CString("Can't create sock") );
        return  FALSE;
    }

    if(!m_pSocket->Connect(lpszAddress, nPort + 700))
    {
        this->m_status_window.SetWindowText( (LPCTSTR) CString("Failed to connect") );
        delete m_pSocket;
        m_pSocket = NULL;
        return FALSE;
```

```
    }
    return TRUE;
}


//
// StartApplication sets the connection parameters and
// calls ConnectSocket. If the connection fails ensure
// that the Control Center (which acts as the Server )
// is running.
//

BOOL CPDABotDlg::StartApplication()
{
    m_strHandle="7";
    m_hostname.GetWindowText(m_strServer);
    m_nChannel=7;

    if (ConnectSocket(m_strHandle, m_strServer, m_nChannel))
            return TRUE;
    else
    {
        this->m_status_window.SetWindowText( (LPCTSTR) CString("Connection Failed ") );
        return FALSE;
    }
}


//
// ReadPDAData() Is called when the CESocket signals that
// data has arrived from the Command Center. The data is
// a string indicating that the connection was successful
// or a Motion command that will be relayed to PDA Robots
// body.
//

void CPDABotDlg::ReadPDAData()
{


    CString status_message;
    char szMessage[512];
    static int initialized;
    u_long numbytes;
    int ret;

    //
    // Ensure that we won't be blocked waiting here on the call
    // to read the data.
    //

    ret = ioctlsocket ((SOCKET) m_pSocket, FIONREAD, &numbytes);

    if( (ret == 0) && (numbytes > 0) )
```

191

```
{
    //
    // Receive the data from Command Center
    //

    ret = recv ( (SOCKET) m_pSocket, &szMessage[0], numbytes, 0);
}
else{
    return;
}

//
// Set the status to the last command so it can be displayed
// in the status edit box named m_status_window and used in the
// OnTimer() function to relay the commands to PDA Robot
//

LastStatus = szMessage;

//
// Note: The Timer was started when the Wireless link was
// Enabled and the data received is interpreted when the timer
// goes off.
//


}


void CPDABotDlg::OnTimer(UINT nIDEvent)
{
    m_status_window.SetWindowText( (LPCTSTR) LastStatus);


    if(LastStatus == "SUCCESS" )
    {
        //
        // We have connected to the Command Center
        //

        m_status_window.SetWindowText( (LPCTSTR) CString("Connect Infrared") );

    }else if(LastStatus == "FORWARD"){

        //
        // Instruct PDA Robot to move Forward via the IR Socket if
        // an IrDA link has been established
        //

        if( m_bIrDAConnected )
        {
            send( Infrared_Socket, (const char *) "be", 2, MSG_DONTROUTE);
```

```
        }

}else if(LastStatus == "REVERSE" ){

    //
    // Instruct PDA Robot to move Reverse via the IR Socket if
    // an IrDA link has been established
    //

    if( m_blrDAConnected )
    {
        send( Infrared_Socket, (const char *) "ad", 2, MSG_DONTROUTE);
    }

}else if(LastStatus == "LEFT" ){

    //
    // Instruct PDA Robot to move Left via the IR Socket if
    // an IrDA link has been established
    //

    if( m_blrDAConnected )
    {
        send( Infrared_Socket, (const char *) "ae", 2, MSG_DONTROUTE);
    }

}else if(LastStatus == "RIGHT" ){

    //
    // Instruct PDA Robot to move Left via the IR Socket if
    // an IrDA link has been established
    //

    if( m_blrDAConnected )
    {
        send( Infrared_Socket, (const char *) "bd", 2, MSG_DONTROUTE);
    }

}else if(LastStatus == "STOP" ){

    //
    // Instruct PDA Robot to Stop via the IR Socket if
    // an IrDA link has been established
    //

    if( m_blrDAConnected )
    {
        send( Infrared_Socket, (const char *) "cf", 2, MSG_DONTROUTE);
    }

}else{
    //
```

```
    // couldn't log on or we received some bad data
    //
  }


    CDialog::OnTimer(nIDEvent);
}
```

Once the wireless connection to the command center (PC) and the infrared connection has been established, we can now control the PDA Robot remotely, seeing through the wireless camera (see **Figure 9.7**).

Figure 9.7

PDA with wireless card.

10

The PDA Robotics Command Center

The command center runs on a Windows 95 or better operating system PC that is connected to the wireless network through a Network Everywhere Cable/DSL Router. It has a Video Capture Card connected to an X10 wireless video receiver.

From the command center, users can control PDA Robot remotely. It can detect motion, as well as save and send images via file transfer protocol (FTP) or simple mail transfer protocol (SMTP). When the application starts, it listens for a connection from the PDA that is controlling PDA Robot. When PDA Robot successfully logs in, users can begin controlling the craft remotely, looking through its eyes. **Figure 10.1** shows the main screen of the command center.

The Video Link

The following program is using the Video for Windows application programming interface (API) provided by Microsoft. A window is created in the dialog, with the Dialog window being the parent, and it is registered as the video window.

```
void CBeamDlg::OnInitializeDriver()
{

    //
    // Display the video source window that allows the user to select the input
```

Figure 10.1

Command center.

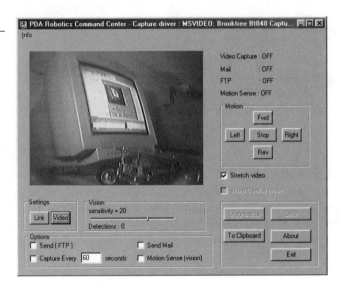

```
//

BOOL code = capDlgVideoSource(h_capwin);

//
// Create the Capture window
//

h_capwin = ::capCreateCaptureWindow("PDABot Video",
        WS_CHILD|WS_CLIPSIBLINGS|WS_VISIBLE|WS_EX_DLGMODALFRAME ,
        20,
        20
        ,320,
        240,this->m_hWnd,0
        );

//
// Hook into the video driver. Check up to 10 and use the first one encountered
//

for( int j=0; j<10;j++){
    code = capDriverConnect(h_capwin, j);
    if(code){
        // Select the video source
        code = capDlgVideoSource(h_capwin);
        // Set the preview rate
        code = capPreviewRate(h_capwin, 100);
        // turn video previewing on
        code = capPreview(h_capwin, TRUE );
        break;
```

```
    }else{
        //
        // No Driver Detected

    }
}

m_stretch_video.SetCheck(1);

//
// Add the name of the video driver to the title bar of the window
//

code = capDriverGetName( h_capwin, &szName[0], wSize );
if( code == TRUE )
{
    title_text += " - Capture driver : ";
    title_text += szName;
    this->SetWindowText(title_text);
}

}
```

The video should now be displayed on the main window, as shown in **Figure 10.1**. If the user has selected motion detection, stream the video into a callback function, checking to see if anything has changed between the first image that was stored in memory and the current frame.

Motion Detection

```
void CBeamDlg::OnCheckVision()
{

    if( m_check_vision.GetCheck() == 1)
    {
        m_radar_stat.SetWindowText("Motion Sense : ON");

        //
        // Disable stretch feature... the stretch requires significant processor cycles
        //

        if( m_stretch_video.GetCheck()  == 1 )
        {
            m_stretch_video.SetCheck(0);
            capPreviewScale(h_capwin, FALSE);
            m_stretch_video.EnableWindow(FALSE);
            this->InvalidateRect(NULL, TRUE);
        }
```

```
//
// Activate the motion detection
//

CAPTUREPARMS cap_params;
long struct_size = sizeof( CAPTUREPARMS );

capCaptureGetSetup( h_capwin, &cap_params, struct_size);
cap_params.fLimitEnabled = FALSE;
cap_params.fAbortLeftMouse = FALSE;
cap_params.fAbortRightMouse = FALSE;
cap_params.wTimeLimit = 3600; // reset after 60 second test
cap_params.fYield = TRUE;

//
// Throw away the audio for now ... will use same algorithm
// for sound detection
//

cap_params.fCaptureAudio = FALSE;
cap_params.wNumVideoRequested = 1;
capCaptureSetSetup( h_capwin, &cap_params, struct_size);

//
// Set the callback to which the vidoe will be stramed
//

BOOL ret = capSetCallbackOnVideoStream( h_capwin, (LPVOID) stream_callback );
capCaptureSequenceNoFile( h_capwin );

    }else{

//
// Unchecked the motion detection
//

m_radar_stat.SetWindowText("Motion Sense : OFF");
capCaptureAbort(h_capwin);
m_stretch_video.EnableWindow(TRUE);
    }
}
```

The video frames will be sent to this callback function where we will compare the last video frame sent to the current

```
LRESULT stream_callback(HWND hWnd, LPVIDEOHDR lpVHdr)
{

    static initialized;
    static DWORD frame_same_total;
    static last_frame_bytes;
```

```
static DWORD trigger_threshold;
static BYTE *last_frame_data;

BYTE *frame_data = (BYTE *) malloc((size_t)lpVHdr->dwBytesUsed);

//
// We will run into a problem if the frame buffer size has
// increased ( user switched the video format settings while
// detecting motion ).so realloc to the correct size.
//

if( !initialized ){
    last_frame_data = (BYTE *) malloc((size_t)lpVHdr->dwBytesUsed);
    last_frame_bytes = (size_t)lpVHdr->dwBytesUsed;
}
else
{
    // Ensure that the bytes used hasn't changed. User may change
    // video settings along the way. Resize our frame buffer

    if( last_frame_bytes != (size_t)lpVHdr->dwBytesUsed )
    {
        // AfxMessageBox( " Reallocating the frame buffer sise !" );
        last_frame_data = (BYTE *) realloc(last_frame_data,
            (size_t)lpVHdr->dwBytesUsed);
        last_frame_bytes = (size_t)lpVHdr->dwBytesUsed;
    }
}


if( (frame_data == NULL ) || ( last_frame_data == NULL ) )
{
    //
    // Frame data couldn't be allocated
    //

    return FALSE;
}

memcpy( frame_data, lpVHdr->lpData, lpVHdr->dwBytesUsed);
if( !initialized )
{
    memcpy( last_frame_data, frame_data, lpVHdr->dwBytesUsed);
    frames_sampled = 0;
    frame_same_total = 0;
    initialized = 1;
}

void *frame_data_start = frame_data;
void *last_frame_data_start = last_frame_data;

//
```

```
// Scan through the frames comparing the last to the new
//

long same_count = 0;
for ( DWORD i = 0; i < lpVHdr->dwBytesUsed; i++ )
{
    if( *frame_data == *last_frame_data )
    {
        same_count++;
    }
    frame_data++;
    last_frame_data++;
}

//
// Reset our pointers or we are wading through deep @#*!
//

frame_data = (BYTE *) frame_data_start;
last_frame_data = (BYTE *) last_frame_data_start;

if( frames_sampled < 5 )
{
    if(frames_sampled > 0 )
    {
        frame_same_total += same_count;
        average_frame_similarity = frame_same_total / frames_sampled;
        trigger_threshold = ( average_frame_similarity / 30 ) *
        global_detection_threshold;
    }

    frames_sampled++;
}


//
// If the slider has been moved recalculate
//

if( recalculate_threshold == 1 )
{
    trigger_threshold = ( average_frame_similarity / 30 ) * global_detection_threshold;
    recalculate_threshold = 0;
}

//
// Note : If sound capture is activated  you can detect the *wave*
// cap_params.fCaptureAudio = TRUE;
//

//
// If we are over the threshold then motion has been detected
```

```
    //

    if( ( same_count < trigger_threshold ) && ( frames_sampled >= 4 ) )
    {

        detected_motion = TRUE;

        //
        // Stop the streaming and grab a frame
        //
        capCaptureAbort(h_capwin);
        capGrabFrame(h_capwin);

        initialized = 0;

        //
        // TODO: ENSURE no mem leakage
        //


        AfxGetMainWnd()->SetTimer(CLEAR_MOTION_DETECT ,50, NULL);

        return TRUE;
    }
    else
    {
        detected_motion = FALSE;
    }

    //
    // Save the last frame
    //

    memcpy( last_frame_data, frame_data, lpVHdr->dwBytesUsed );
    free(frame_data);
    return TRUE;
}
```

When motion is detected, the program will save an image and forward it via FTP or SMTP (mail).

Sending Data Using FTP

```
class CFtp
{

public:

    CFtp();
    ~CFtp();
```

```
        BOOL UpdateFtpFile( CString host,
                            CString user,
                            CString password,
                            CString remote_path,
                            CString filename,
                            CString remote_filename,
                            CString& status_msg );

        BOOL CFtp::TestConnect( CString host,
                                CString user,
                                CString password,
                                INTERNET_PORT port,
                                CString& status_msg);

protected:

        CInternetSession* m_pInetSession; // objects one and only session
        CFtpConnection* m_pFtpConnection; // If you need another create another Cftp

};


CFtp::CFtp()
{

    m_pFtpConnection = NULL;

    // the CInternetSession will not be closed or deleted
    // until the dialog is closed

    CString str;
    if (!str.LoadString(IDS_APPNAME))
        str = _T("AppUnknown");

    m_pInetSession = new CInternetSession(str, 1, PRE_CONFIG_INTERNET_ACCESS);

    // Alert the user if the internet session could
    // not be started and close app
    if (!m_pInetSession)
    {
        AfxMessageBox(IDS_BAD_SESSION, MB_OK);
        OnCancel();
    }


}


// Destructor

CFtp::~CFtp()
```

```
{
    // clean up any objects that are still lying around
    if (m_pFtpConnection != NULL)
    {
        m_pFtpConnection->Close();
        delete m_pFtpConnection;
    }
    if (m_pInetSession != NULL)
    {
        m_pInetSession->Close();
        delete m_pInetSession;
    }
}


// Update our file

BOOL CFtp::UpdateFtpFile(CString host, CString user, CString password, CString
remote_path, CString filename, CString remote_filename, CString& status_msg)
{

    CString strFtpSite;
    CString strServerName;
    CString strObject;
    INTERNET_PORT nPort;
    DWORD dwServiceType;

    if (!AfxParseURL(ftp_host, dwServiceType, strServerName, strObject, nPort))
    {
        // try adding the "ftp://" protocol
        CString strFtpURL = _T("ftp://");
        strFtpURL += host;

        if (!AfxParseURL(strFtpURL, dwServiceType, strServerName, strObject, nPort))
        {
            // AfxMessageBox(IDS_INVALID_URL, MB_OK);
            // m_FtpTreeCtl.PopulateTree();
            status_msg = "Bad URL, please check host name";

            return(FALSE);
        }
    }

    // If the user has provided all the information in the
    // host line then dwServiceType, strServerName, strObject, nPort will
    // be filled in.. but since I've provided edit boxes for each we will use these.

    // Now open an FTP connection to the server
    if ((dwServiceType == INTERNET_SERVICE_FTP) && !strServerName.IsEmpty())
    {
        try
        {
```

```
        m_pFtpConnection = m_pInetSession->GetFtpConnection(strServerName, user,
password, 21 );
        }

    catch (CInternetException* pEx)
    {
        // catch errors from WinINet
        TCHAR szErr[1024];
        if (pEx->GetErrorMessage(szErr, 1024))
            // AfxMessageBox(szErr, MB_OK);
            status_msg = szErr;
        else
            status_msg = szErr;
            //AfxMessageBox(IDS_EXCEPTION, MB_OK);
        pEx->Delete();
        m_pFtpConnection = NULL;
        return(FALSE);
    }

}
else
{
    status_msg = "Bad URL, please check host name";

}

BOOL rcode = m_pFtpConnection->SetCurrentDirectory(remote_path);
if( FALSE == rcode )
{
    status_msg = "Could not goto directory specified. Please re enter";

}

CString strDirName;
rcode = m_pFtpConnection->GetCurrentDirectory(strDirName );

rcode = m_pFtpConnection->PutFile( filename, (LPCTSTR) remote_filename,
        FTP_TRANSFER_TYPE_BINARY, 1 );
if( FALSE == rcode )
{
    status_msg = "Could not update file. Check settings";
    return(FALSE);
}

    return(TRUE);
}

// Test connection

BOOL CFtp::TestConnect(CString host, CString user, CString password, INTERNET_PORT
port, CString& status_msg)
{
```

```
CString strFtpSite;
CString strServerName;
CString strObject;
INTERNET_PORT nPort;
DWORD dwServiceType;

// If the user has provided all the information in the
// host line then dwServiceType, strServerName, strObject, nPort will
// be filled in.. but since I've provided edit boxes
// for each we will use these.

//
// Ensure Valid connection parameters
//

CString diagnostic_msg = "";

if ( host.IsEmpty() )
{
    diagnostic_msg = " check Host ";
}
if ( ( user.IsEmpty()) || (user == "") )
{
    diagnostic_msg = " check Username ";
}

if ( password.IsEmpty() )
{
    diagnostic_msg = " check password ";
}

if ( port < 1 )
{
    diagnostic_msg = " check port ";
}

// Now open an FTP connection to the server
try
{
    m_pFtpConnection = m_pInetSession->GetFtpConnection(host, user, password, port );
}

catch (CInternetException* pEx)
{
    // catch errors from WinINet
    TCHAR szErr[1024];
    if (pEx->GetErrorMessage(szErr, 1024))
    {
        // AfxMessageBox(szErr, MB_OK);
        status_msg += diagnostic_msg;
        status_msg += szErr;
```

205

```
    }else{
        status_msg += diagnostic_msg;
        status_msg = szErr;
    }
        //AfxMessageBox(IDS_EXCEPTION, MB_OK);

    pEx->Delete();
    m_pFtpConnection = NULL;
    return(FALSE);
    }

    return(TRUE);
}
```

The Wireless Data Link

On startup, the command center listens on a socket for a connection request from the PDA controlling PDA Robot. The listening socket is derived from the CCeSocket, as is the socket created on the PDA.

```
#include "stdafx.h"


CListeningSocket::CListeningSocket(CBeamDlg* pDoc)

void CListeningSocket::OnAccept(int nErrorCode)
{
    CSocket::OnAccept(nErrorCode);
    m_pDlg->ProcessPendingAccept();
}


CListeningSocket::~CListeningSocket()
{
}
IMPLEMENT_DYNAMIC(CListeningSocket, CSocket)
```

The Main Dialog window of the command center contains the ClisteningSocket as a member variable. When we receive the request and it is authenticated, we send the string "SUCCESS" back to the PDA. Because the socket is established, we can send commands such as "FORWARD" to the PDA, which sends the corresponding commands to PDA Robot via the infrared Link.

```
CBeamDlg::CBeamDlg(CWnd* pParent /*=NULL*/)
    : CDialog(CBeamDlg::IDD, pParent)
{
    //{{AFX_DATA_INIT(CBeamDlg)
    // NOTE: the ClassWizard will add member initialization here
```

206

```
    //}}AFX_DATA_INIT
}

CBeamDlg::~CBeamDlg()
{
    //m_oAnimateCtrl.Stop();
}

void CBeamDlg::DoDataExchange(CDataExchange* pDX)
{
    CDialog::DoDataExchange(pDX);
    //{{AFX_DATA_MAP(CBeamDlg)
    DDX_Control(pDX, IDC_PROGRESS1, m_WndProgressCtrl);
    //}}AFX_DATA_MAP
}


BEGIN_MESSAGE_MAP(CBeamDlg, CDialog)
//{{AFX_MSG_MAP(CBeamDlg)
ON_WM_TIMER()
ON_WM_CLOSE()
ON_BN_CLICKED(IDC_BUTTON1, OnDetailsClick)
//}}AFX_MSG_MAP
END_MESSAGE_MAP()

// CBeamDlg message handlers

BOOL CBeamDlg::OnInitDialog()
{
    CDialog::OnInitDialog();

    m_pClientSocket = new CClientSocket(this);

    m_pSocket = new CListeningSocket(this);
    if (m_pSocket->Create( 707 ))
    {
        if (m_pSocket->Listen())
        {
            return TRUE;
        }
    }

    return TRUE;
}


void CBeamDlg::ProcessPendingAccept()
{
    char szHost[25];
    if (m_pSocket->Accept(*m_pClientSocket))
    {
        m_pClientSocket->Receive(szHost,25,0);
```

```
        m_WndProgressCtrl.StepIt();

        m_csHost=szHost;
    }
}


void CBeamDlg::ProcessPendingRead()
{
    char szUserName[255];
    char szPassword[255];

    m_pClientSocket->Receive(szUserName,255,0);
    m_pClientSocket->Receive(szPassword,255,0);

    m_WndProgressCtrl.StepIt();
    GetDlgItem(IDC_STATUS_STATIC)->SetWindowText("Authenticating..");

    if(!CheckForAuthentication(szUserName,szPassword))
    {
        m_pClientSocket->Send("ERROR",255,0);
        GetDlgItem(IDC_STATUS_STATIC)->SetWindowText("Bad user name or password..");
    }
    else
    {

        //Recv the instant message here

        m_pClientSocket->Send("SUCCESS",255,0);
        ///play sound to inform the user
        PlaySound("wireless.wav",NULL,SND_FILENAME);
    }
}

BOOL CBeamDlg::CheckForAuthentication(char * pszUserName,char *pszPassword)
{
    WCHAR szUserName[100];
    WCHAR szPassWord[100];

    //
    //Convert to unicode
    //

    MultiByteToWideChar(CP_ACP, 0, pszUserName,
        strlen(pszUserName)+1, szUserName,
        sizeof(szUserName)/sizeof(szUserName[0]) );

    //
    //Convert to unicode
    //

    MultiByteToWideChar(CP_ACP, 0, pszPassword,
```

```
        strlen(pszPassword)+1, szPassWord,
        sizeof(szPassWord)/sizeof(szPassWord[0]) );

    //
    // Determine if the password is correct by changing it
    // to the same.
    //

    int nStatus= NetUserChangePassword(NULL,
        szUserName,
        szPassWord,
        szPassWord);
    if (nStatus == NERR_Success){
        return TRUE;
    }else{
        return FALSE;
    }


}
```

Please visit www.pda-robotics.com to download this program.

11

Infinitely Expandable

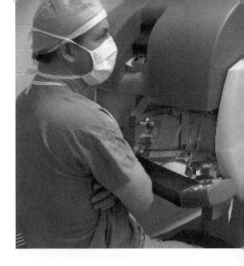

PDA Robot is a fusion of the latest technologies on all fronts. The way technology evolves, this may not be true for long. I hope that users take from this project not only knowledge of the technology, but the realization that with a little research and a Web browser, users can find a solution to any problem.

PDA Robot can be easily expanded to use the wide range of add-on technology available, such as a global positioning system (GPS) card. This chapter lists a number of cards and pieces of equipment that could be used with this project, and concludes with a great piece of equipment used for telesurgery.

Global Positioning System

These devices allow users to get an exact position on where they are located. This means that users can program PDA Robot to autonomously go to any location on the earth or navigate with the aid of a long-range wireless video transmitter anywhere in the city. I will have to do my own calculations for the moon mission by triangulation off three transmitter beacons, though. If you buy one, be sure that you can get the position information through an application programming interface (API) that the manufacturer provides. I won't buy one if I can't write a program to access the data.

Pocket CoPilot 3.0 GPS Jacket Edition: PCP-V3-PAQJ2

The iPAQ Pocket PC-based navigation solution guides users safely and intuitively, using detailed voice directions (full street names). It has the following features:

- Seamless nationwide routing from anywhere to anywhere.

- Fastest full route automatic recalculation.

- Traffic congestion detour feature.

- Superior CoPilot GPS jacket with 4x faster acquisition; 50% reduction in power consumption; sleek, ultra-thin design; and integrated power port, battery, and CF slot.

- Vehicle mount, power adapter, and complete nationwide map data.

- BMW Mini certified accessory option for Europe.

Users can turn their iPAQ into an amazing GPS navigation system with Pocket CoPilot (see **Figure 11.1**). This exciting new technology gets users precisely where they need to be, with directions to any address nationwide.

And with dynamic voice navigation and route guidance technology, Pocket CoPilot not only shows users where to go, it verbally guides them to their destination in real-time, with audible text-to-speech directions. Yes, you will hear, "1.3 miles ahead, turn left on South Street." If you miss a turn or get off track, Pocket CoPilot automatically reroutes you (**Figure 11.1**).

The TeleType GPS

I like the GPS PCMCIA Receiver Card teletype card, because it fits right into my expansion pack.

The Wireless PCMCIA GPS receiver has been designed especially for use with the popular Compaq iPAQ Pocket PC. The combination of the GPS receiver and the TeleType GPS software allows travelers to navigate worldwide via land, air, and water using a completely integrated device, eliminating cumbersome wires. The TeleType Wireless PCMCIA package includes the TeleType GPS software and street-level maps for the entire United States allowing real-time position to be accurately shown (see **Figure 11.2**).

Figure 11.1

CoPilot.

Figure 11.2

TeleType.

Symbol SPS 3000 Bar Code Scanner Expansion Pack

Users can increase the effectiveness of their iPAQ Pocket PC with powerful data capturing capabilities. Data capture through bar code scanning is more accurate and significantly improves productivity, creating a dynamic business tool for your workforce. The Symbol SPS 3000 enables one-dimensional bar code scanning, and is available in the following two feature configurations:

- Bar code scanning only:

  - Very low power consumption.

- Bar code scanning with integrated wireless local area network (WLAN):

  - Integrated 802.11b WLAN (see **Figure 11.3**).

  - Internal battery to power the WLAN radio.

Symbol Technologies, Inc. will provide service and support warranty. More information, available from Symbol, includes SDK, technical specs and driver downloads (**Figure 11.3**).

It's the NEX best thing... the Compaq iPAQ Pocket PC becomes a digital camera—ideal for business and personal use. Users can quickly upload photos to their desktop or laptop and access full-color photograph images (24-bit SVGA; 800 × 600).

Figure 11.3

Symbol scanner
integrated 802.11b.

Sierra Wireless AirCard 555

With the AirCard 555, users will also be able to make voice calls and send SMS (two-way messaging) messages.

With the dual-band 1X Sierra Wireless AirCard 555, users will experience the following:

*   Faster speeds: up to 86 kb/s, making it more efficient to access your time-sensitive information.

*   Always being connected: allows users to maximize productivity by becoming dormant when data are not actively being transferred through their wireless connection, yet still maintains a virtual connection to the network. This will free up resources, allowing users to multitask by sending or receiving voice calls or text messages on their device. When users are ready to resume their data session, they can re-engage the network immediately—there's no need to dial in again, and no waiting. Instant-on Bell Mobility acts as the Internet service provider (ISP). Users can connect in seconds to the information they need.

The Sierra Wireless AirCard 555 lets users access their information simply by sliding it into the PC Card slot in their laptop computer or handheld device (see **Figure 11.4**). Coupled with the easy-to-install software included in the kit, and a Bell Mobility 1X Data plan, the AirCard 555 transforms the device into a complete wireless business solution.

Figure 11.4

AirCard.

Telesurgery

Dr. Louis Kavoussi uses the Internet to lend expertise to operating rooms all over the world.

With the help of the Internet and telecommunications technology, this doctor in Baltimore, Maryland can operate on patients all over the world without leaving his home office. Working from home using a PC and four ISDN lines, Dr. Kavoussi of Johns Hopkins Bayview Medical Center controls robotic surgical tools and cameras remotely, and can transmit and view images in real time.

During surgery, Kavoussi can view either the operating room or inside the patient. He can also give surgeons written assistance and operate a device that burns and seals tissue, as well as control robots that hold cameras or place needles in the patient's body.

"Our applications have been used specifically for what's called minimal invasive surgery," said Kavoussi. "Examples of that are laparoscopies, putting a little tube in the stomach to look around; arthroscopy, looking at knee joints; and thoracoscopy, looking at the chest."

Operations of the Future

With the help of high-speed data lines and advanced robotics, surgeons will eventually be able to perform and complete operations remotely from anywhere in the world.

"There is no doubt in my mind that this is the way surgical care is performed in the future," Kavoussi said.

Doctors in New York took telemedicine one step further when they used a dedicated fiber-optic line and a remote-control robot to remove the gall bladder of a patient in an operating room in France—more than 4,000 miles away.

Telesurgery may also be employed during future space exploration. Traveling to Mars and back may take three or more years, during which time astronauts may need access to medical and surgical care.

The da Vinci Robotic System. The da Vinci Surgical System is integral to the operating room and supports the entire surgical team. The system consists of a surgeon console, patient-side cart, instruments and image processing equipment.

- *The Surgeon Console:* The surgeon operates at the console using masters (that replicate surgery motions) and the high-performance vision system that is controlled using foot pedals and displays in the same orientation of open surgery. Using the da Vinci Surgical System, the surgeon operates while seated comfortably at a console viewing a 3-D image of the surgical field. The surgeon's fingers grasp the master controls below the display, with wrists naturally positioned relative to his or her eyes. This technology seamlessly translates the surgeon's movements into precise, real-time movements of the surgical instruments inside the patient.

- *Patient-Side cart:* The patient-side cart provides the two robotic arms and one endoscope arm that execute the surgeon's commands. The laparoscopic arms pivot at the 1 cm operating port, eliminating the use of the patient's body wall for leverage, minimizing tissue and nerve damage. Supporting surgical team members install the correct instruments, prepare the 1 cm port in the patient, and supervise the laparoscopic arms and tools being utilized.

- *EndoWrist Instruments:* A full range of instruments is provided to support the surgeon while operating. The instruments are designed with seven degrees of motion to mimic the dexterity of the human wrist. Each instrument has a specific surgical mission such as clamping, suturing, and tissue manipulation. Quick-release levers speed instrument changes during surgical procedures.

- *Insite High Resolution 3-D Endoscope and Image Processing Equipment:* Provides the true to life 3-D images of the operative field. Operating images are enhanced, refined, and optimized using image synchronizers, high-intensity illuminators, and camera control units.

The da Vinci Surgical System is the only commercially available technology that can provide the surgeon with the intuitive control, range of motion, fine tissue manipulation capability, and 3-D visualization characteristic of open surgery, while simultaneously allowing the surgeon to work through small ports of minimally invasive surgery (see **Figure 11.5**).

Figure 11.5

da Vinci system.

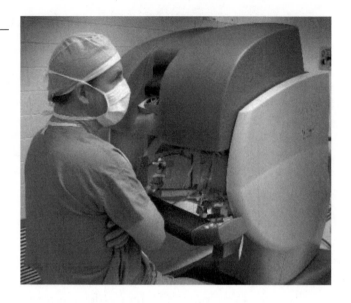

Robotic Heart Surgery: Making Repairs without Lifting the Hood. In the United States, open-heart surgery was performed without opening the chest, in more than a dozen patients. Researchers reported preliminary results at the American Heart Association's Scientific Sessions in 2002.

In this procedure, surgeons remotely maneuver robotic arms from a seat in front of a console away from the patient. Instead of opening the chest and cutting the skin and muscle to view the area, surgeons make four holes (8 to 15 mm each), through which robotic arms are inserted. The robotic arms include one with a camera-like device to transmit the image to the console. The other arms are fitted with operating instruments.

Surgeons used this new procedure to successfully repair the hearts of patients with atrial septal defect (ASD) or patent foramen ovale—conditions in which people are born with an opening between the heart's two upper chambers. This opening allows some blood from the left atrium to return to the right atrium, instead of flowing through the left ventricle, out the aorta, and to the body. It is repaired either by plugging the hole with a patch or suturing the hole closed.

Open-heart surgery traditionally requires that surgeons make a foot-long chest incision to cut patients' breastbones in half. "We wanted to know if it was possible to operate inside the hearts of these patients without making any incisions," says Mehmet Oz, M.D., director of the Heart Institute at Columbia-Presbyterian Medical Center in New York. "Not only did we show that the operation is feasible, but we demonstrated it in more than a dozen patients."

During 12 months, 15 patients (ages 22 to 68) underwent ASD repair using the robotic technology, called the da Vinci system described in the preceding section. "Although the equipment is costly, this is definitely part of the future," says Michael Argenziano, M.D., lead author of the study and director of robotic cardiac surgery at Columbia-Presbyterian. "Patients are going to insist on it despite the expense because it's cosmetically superior and allows for much faster recovery. For certain procedures, like the ASD repair, it's already proving to be a worthy alternative to conventional surgery."

The researchers found that robot-assisted endoscopic heart surgery takes a little longer than the traditional technique, but that might be attributable to the learning curve necessary to use the new approach. The heart was stopped for 34 minutes on average, versus about 20 for traditional surgery. The time needed on a cardiopulmonary bypass machine was also slightly longer.

Patients in the study had no major complications. In 14 cases, imaging tests confirmed that the defect had been successfully closed. One patient required a repair five days later. Surgeons did this through a three-inch incision (a mini-thoracotomy). The average length of stay in the intensive careunit was 18 hours, which is about the same as for the traditional approach. The average hospital stay was three days—two to four days shorter than for a traditional operation.

"The primary advantages of this minimally invasive surgery are faster patient recovery, less pain, and dramatically less scarring than traditional open-heart surgery," Argenziano says. Patients return to work and normal activity about 50 percent faster than those who have the open procedure, he says. Quality-of-life measures also revealed the robotically treated patients had improved social functioning and less pain compared to patients undergoing traditional surgical approaches. Doctors are also using the robotic technology to repair mitral valve defects through incisions in the side of the chest.

"What makes the totally endoscopic ASD repair a significant advance is that it is the first closed-chest open-heart procedure," Argenziano says.

219

Argenziano is also principal investigator of several Food and Drug Administration-sanctioned trials of robotic cardiac surgery including one in which it is used for closed-chest coronary artery bypass graft surgery (CABG). In early 2002, the Columbia team performed the first totally endoscopic CABG in the United States.

"We have wonderful surgical cures for heart disease, in that they're very effective and long-lasting," Oz says. "However, they're also very traumatic. So, we're evaluating a technology that might provide us with the same wonderful results without the trauma."

Several facilities nationwide offer the da Vinci technology, and researchers at approximately four other centers have been specifically trained to perform ASD closure, the researchers say.

Index

Note: Boldface numbers indicate illustrations and tables.

About the Author

Doug Williams is a software designer for Agfa Healthcare. A resident of Ontario, Canada, he has worked in the computer industry for nearly 10 years, specializing in radar systems control, medical imaging software, and electronic interface technologies.